ARCHIVING MACHINES

The Information Society Series

Laura DeNardis and Michael Zimmer, Series Editors

Interfaces on Trial 2.0, Jonathan Band and Masanobu Katoh

Opening Standards: The Global Politics of Interoperability, Laura DeNardis, editor

The Reputation Society: How Online Opinions Are Reshaping the Offline World, Hassan Masum and Mark Tovey, editors

The Digital Rights Movement: The Role of Technology in Subverting Digital Copyright, Hector Postigo

Technologies of Choice? ICTs, Development, and the Capabilities Approach, Dorothea Kleine

Pirate Politics: The New Information Policy Contests, Patrick Burkart

After Access: The Mobile Internet and Inclusion in the Developing World, Jonathan Donner

The World Made Meme: Public Conversations and Participatory Media, Ryan Milner

The End of Ownership: Personal Property in the Digital Economy, Aaron Perzanowski and Jason Schultz

Digital Countercultures and the Struggle for Community, Jessica Lingel

Cyberbullying Policies of Social Media Companies: Toward Digital Dignity, Tijana Milosevic

Authors, Users and Pirates: Copyright Law and Subjectivity, James Meese

Weaving the Dark Web, Robert W. Gehl

Hacker States, Luca Follis and Adam Fish

The Politics of Dating Apps: Gender, Sexuality, and Emergent Publics in Urban China, Lik Sam Chan

Technology of the Oppressed, David Nemer

The Political Lives of Information: Information and the Production of Development in India, Janaki Srinivasan

Living with Algorithms: Agency and User Culture in Costa Rica, Ignacio Siles

The Secret Life of Data: Navigating Hype and Uncertainty in the Age of Algorithmic Surveillance, Aram Sinnreich and Jesse Gilbert

Politics Recoded: The Infrastructural Organizing of Code for America, Aure Schrock

Archiving Machines: From Punch Cards to Platforms, Amelia Acker

ARCHIVING MACHINES

FROM PUNCH CARDS TO PLATFORMS

AMELIA ACKER

The MIT Press
Cambridge, Massachusetts
London, England

The MIT Press
Massachusetts Institute of Technology
77 Massachusetts Avenue, Cambridge, MA 02139
mitpress.mit.edu

The MIT Press would like to thank the anonymous peer reviewers who provided comments on drafts of this book. The generous work of academic experts is essential for establishing the authority and quality of our publications. We acknowledge with gratitude the contributions of these otherwise uncredited readers.

This book was set in Bembo Book MT Pro by Westchester Publishing Services. Printed and bound in the United States of America.

Library of Congress Cataloging-in-Publication Data

Names: Acker, Amelia author
Title: Archiving machines : from punch cards to platforms / Amelia Acker.
Description: Cambridge, Massachusetts : The MIT Press, 2025. | Series: The information society series | Includes bibliographical references and index.
Identifiers: LCCN 2024060777 (print) | LCCN 2024060778 (ebook) | ISBN 9780262553247 paperback | ISBN 9780262384568 pdf | ISBN 9780262384575 epub
Subjects: LCSH: Information storage and retrieval systems—Electronic records—History | Information storage and retrieval systems—Archival materials—History | Digital preservation—History | Archives—Technological innovations—History
Classification: LCC CD974.4 A35 2025 (print) | LCC CD974.4 (ebook) | DDC 025.8/409—dc23/eng/20250415
LC record available at https://lccn.loc.gov/2024060777
LC ebook record available at https://lccn.loc.gov/2024060778

10 9 8 7 6 5 4 3 2

EU Authorised Representative: Easy Access System Europe, Mustamäe tee 50, 10621 Tallinn, Estonia | Email: gpsr.requests@easproject.com

CONTENTS

ACKNOWLEDGMENTS

Writing a book is like building a bridge. It needs different kinds of support to complete it, and there are so many kinds of support that I wish to acknowledge here.

First, I would like to acknowledge the many institutions that supported the writing of this book with research funding for the time to travel and access archives. I would particularly like to thank the Institute for Museum and Library Services, which funded a career grant on exploring digital preservation in mobile and social media platforms (award RE-07-18-0008-18). I would like to acknowledge the National Science Foundation for funding a collaborative grant with Megan Finn on scientific data management and the afterlives of data (award SES-2020183). In 2018, I was a fellow at the Media Archaeology Lab, where I spent wonderful, deep research week immersed in their collections of handheld digital assistants and portable computers in Boulder, Colorado. The Office of the Vice President for Research, Scholarship and Creative Endeavors at the University of Texas at Austin provided generous support for editing and production.

My closest research collaborators in my early career have been Megan Finn, Andrew Iliadis, Joan Donovan, and my dear Sistra. Each have been steadfast supporters of this book since the early days of imagining what it would be. I owe a great debt to my incredible book club: Tanya Clement, Melanie Feinberg, and Heather Houser. As my first readers, this group read every chapter, from early drafts onward, they offered steadfast encouragement and helpful feedback month after month. I hope we keep writing and reading books together!

From 2016 to 2025, I had the honor of working at the University of Texas at Austin. I am grateful for the inspiring colleagues and students whom I met there. I would especially like to thank folks from the School of Information, particularly James Howison, Jakki Bailey, Min Kyung Lee, and Angela D. R. Smith. Leadership at the iSchool was especially supportive of me writing this monograph. Some of my closest colleagues at UT have influenced me to write boldly and with enthusiasm. In this regard, I am very grateful to Samantha Shorey, Edgar Gómez Cruz, Iván Chaar López, and Soo Young Rieh.

In the last months of writing this book, one of my dearest coworkers, Ciaran Trace, was taken from us too soon. Ciaran was an important example to me as an educator and an academic community member. She was an indefatigable colleague, a passionate archival scholar, and a spirited teacher. Among the many colleagues with whom I've had the pleasure of working, Ciaran's imaginative skill and joyfulness always inspired me, and I have tried to honor her influence and significant contributions to archival scholarship with some of the concepts developed here.

In 2023, I was fortunate to host a book workshop that substantially shaped the connective themes across the book. I am grateful to Andrew Russell, Megan Ankerson, Kevin Driscoll, and Stacy E. Wood for giving their time, attention, and support as initial readers of the first full draft of the manuscript. I'm especially appreciative of my friend Sarah Brayne, who gave inspiration to ask for the support to organize a workshop, and to Eric T. Meyer for supporting it with the Mary R. Boyvey Fellowship. I believe that this workshop was the best gift to me, as it helped me build connections across computing eras and archival issues by envisioning my readers.

I have had the great honor of working with, supervising, and mentoring distinguished early career scholars through their doctoral studies during the writing of this book. I am especially grateful to Yubing Tian, Hana Frluckaj, and Krishna Akhil Kumar Adavi. Lucy Flamm and MiriYam Judd were both incredible research assistants for much of the physical artifact research. My students remind me of my own wonderful time as a graduate student and the community that sustained me. There are many fellow travelers from graduate school that encouraged me and supported earlier threads of this work. Many thanks to David J. Kim, Roderic Crooks, Matt Mayernik, Tonia Sutherland, Matt Burton, Adam Kriesberg, Andrea Thomer, Morgan Currie, Britt Paris, Anna Lauren Hoffman, Nicholas Proferes, Shawn Walker, Lana Swartz, Lilly Irani, Patrick Keilty, and Andrew J. Lau.

Early roots of this book began when I was pursuing doctoral training after working as an archivist processing mid-century collections. During that time, there were many important professors and mentors such as

Anne J. Gilliland, Johanna Drucker, Sharon Traweek, Deborah Estrin, Paul Dourish, and Ronald E. Day, each of whom gave me encouragement that spurred my confidence as a young information researcher. I am indebted to those teachers who taught me to read closely and to advocate for my own work to others too. I am especially grateful to Gregory Leazer and Christine Borgman for their creative power in the classroom. All of these teachers continue to inspire me.

Throughout the book, I discuss historic computing machines ranging from mainframe computers and minicomputers to personal digital assistants and early smartphones. I also discuss these machines' external storage formats, such as magnetic tape or Subscriber Identity Module (SIM) cards. Unless otherwise noted, these artifacts were studied during my fieldwork at the Media Archaeology Lab at CU Boulder, the Living Computers: Museum + Labs, and the Computer History Museum archives, and the Bancroft Library at UC Berkeley from 2018 to 2022. Fieldwork with machines typically involved powering them on, mounting a program or operating system, and interacting with them. Where possible, I endeavored to create, save, store, or transfer files to relevant formats. Along the way, there have been many librarians, archivists, and preservationists who have helped me with this work, I would especially like to thank Cynde Moya, Penny Ahlstrand, Emily Matteson, and Meryl Brodsky for their enthusiasm and research support.

My true blue Texas friends, Mary Alice Foster, Daniel Carter, Selene Arrazolo, and Jessica Meyerson, all gave me good laughs and long walks when I needed them most. Thanks to our lovely neighbors, Lesley Taylor and Jett Butler, for the beauty and balance they cultivate on Cliff Street. I always carry my incredible California friends in my heart. They are Jorge Rivas, Charchi Stinson, Elizabeth Birkenbuel, Devina Sindhu, and Emily Virgil. I also want to thank the Kelly family, the Acker family, and my parents, Debbie and Stephen.

My ardor for this project was sustained by Jane Kelly, who listened, asked questions, fed me, and gave me encouragement from early drafts to the project's completion. Jane lifted me up with her faithful support in the book and kept me smiling. Finally, I want to thank my dog and best friend, Charlotte Brontë, who sat and napped beside me through it all.

1

ARCHIVING DATA

Each morning, like so many other people, I start writing by firing up my word processor. Recently, the executable program that I use on my laptop became an online application. It is now just one app in a suite of software tools that my institution licenses. When I shut down my computer for the day, I instinctively go to save my file—but there is no longer a "save" option. My online word processor instead has a message in its menu that reads "Where's the Save Button? There's no Save button because we're automatically saving your document." The banality of this dialog response always intrigues me because it elicits more questions that the drop-down menu has conspicuously left unanswered. Why is the document saved automatically? Where is it saved? How is it stored? Who decides this for me? What are the stakes of saving and accessing information automatically by platforms that are managed in the cloud rather than as a file on my laptop's hard drive?

This book argues that the answer to these questions lies in the evolution of archiving data from storing files to managing digital information that is now born networked and automatically saved. I refer to today's archives as "born networked" because corporate data collection is now seamlessly integrated into the networked technologies that we use most. Platforms, rather than public institutions, manage and control much of our data. In presenting a history of mid-century data management and the rise of asymmetrical access to data stores, this book shows how "archive" became a verb, infused with the power to control what information we can save and the means to access it.

Processes for archiving data and managing access to data collections are the beating heart of information management systems that drive new communication technologies, such as cell phones, apps, and social networking platforms. Indeed, contemporary data-driven technologies are premised on the ability to store, preserve, and manage access to data. Over many decades, all contemporary information infrastructures, ranging from financial systems to supply chains, have devoted energy and resources to managing data access across networks. Whether making a mobile payment or tracking vaccine distribution, these common information processes rely on data management—practices of collecting, organizing, and accessing stores of data over time and place. But contemporary data management processes differ from earlier information management approaches to access because they ingest bits of data and interpolate them to more mutable formats for later reuse.[1] Unlike formatted information found in documents and records, digital data is easier to slice up, recombine, blend, and link together into new streams of resources.

Despite their ubiquity, data archives are increasingly hard for people to find, know, or even envision because their downstream access points are increasingly distributed and invisible. Access is typically distanced from users once data is created in networks. Much like the automated processes that have replaced the "save" button for my word processing software, the means of accessing data that we create are increasingly unknown to end users. In this book, a brief history of mid-century data management shows how data became distanced from their original contexts of creation and creators, eventually becoming invisible through processes of data management and archiving with computing machines. Large corporations and platforms that collect and manage collections now administer access and control what can ultimately be known about our data-driven society. The "archiving machines" of this book's title have transformed much of the actions now taken to preserve, manage, and access evidence of our digital culture. But archiving machines are not restricted to traditional, institutional archives, and the very invisible enclosure of archiving data processes obscures their power to determine what is added to our digital cultural memory.

BORN NETWORKED DATA

We live in a moment when creating digital data has become necessary to exist in society. In exploring the history of data management as collections of data became digital, this book charts how data is now *born networked*. These traces of digital data appear to be seamlessly born into networked information infrastructures. Our mobile apps allow us to hail rides and order prescriptions, social media posts help us connect with others, and reading digital books enables us to understand the ideas of others. All these activities create streams of data born in networks—digital traces that have demanded new forms of storage, exchange, and transmission. Collectively, I argue that these processes have given rise to a new form of archival power found in data management that distances people from the data they produce. In distancing data from creators, there is power in managing who can approach, and provide access to data once it is formatted and stored. Archives have always asserted power in controlling access to information. But more and more private corporations, rather than the state and the public, now control access to information in data archives made up of these born networked traces.

The beginnings of this shift in power can be found in computing technologies that enable new caches of data to be stored and managed with what I call "archiving machines." Today, these machines provide access to data as evidence as earlier archives provided access to information and official records. However, unlike earlier eras of archives and information management that enabled access through recorded documents such as files and punch cards, today's data collections are mobile, mutable, and easily reconfigured because of these archiving machines. This flexibility has been made possible because data are now digital, stored in cloud architectures, and born in networks. Scholars have traced this power shift to datafication, with cloud computing, mobile phone apps, platforms, and software services originating in the late 1990s and early 2000s.[2] However, in this book, I argue that the transition from analog data archives to digitally born networked data occurred much earlier. The impact of this transition on archives, data management, and access to evidence has not been closely examined by critical data studies scholars. Although most accounts of datafication focus on the design and implementation of new internet technology, particularly tech adoption contexts where interactions result in data enclosed by corporate platforms, I contend that the history of archiving data, with computing technologies and proposals from early information professionals responding to the storage and preservation challenges of new forms of data, begins in government and academic contexts before the rise of the commercial, public internet.

Beginning in the 1960s, computers started to be designed with memory and storage structures that allowed for more complex capabilities to process larger amounts of information.[3] With these new architectures, computers could be shared by partitioning resources, or "time-share" across multiple users and tasks, processing routines faster thanks to expanded memory cores. The rise of more compact computer systems and data storage formats such as magnetic tape made possible the ability to generate, compute, and store data at larger scales. Using operating systems, protocols, and code stored on tapes, then mailed and hand-delivered to university research labs throughout the US, many of these machines would eventually be linked as nodes in a network called the ARPANET. Their early storage and distribution practices would then lead to a culture of sharing, copying, and accessing data across distances via networked computers, now common to internet cultures today.[4] These advances in data management and data transmission set the stage for future shifts in information technology that would eventually disempower the very public institutions that gave rise to time-sharing, open architecture, and the internet.

The earliest data archives leveraged both analog and mechanical data processes to manage collections of punch cards and paper records stored in filing cabinets. Beginning in the 1940s, data archives, also known as "data banks," were established just as more affordable, external storage appeared. These new digital memory formats that took up less space on shelves and were easier for people to handle and for machines to process. Slowly, standardized digital storage formats began to replace the analog information management systems, which had been paper-driven with hanging files, and punch cards for hulking mainframe computers.[5] Preserving and representing data caches for quick retrieval and access at scale further grew during the 1970s–1990s, forming the base of consumer cloud services and modern data centers that are now used in all digital and networked transactions supported by the internet. As more data was created in online networks and public information infrastructures, it became subject to collection and tracking, becoming part of not only common computer utilities and software services but surveillance programs as well.

Computing architectures for distributed storage and faster processing technologies unleashed the potential for collecting data at incredible scale, ushering in an era of "big data." However, this power came at a cost of control in managing data for future use and how access is designed. Automating computing processes such as saving files to disks reduced the need for technical expertise from computer users to store and save their information, opening a distance between people and the data they generated.

Today, the tech industry regards nearly all human behaviors—from our steps to our sleeping patterns—to be ripe for profit.[6] It is presumed that nearly all human activity can be represented as data and metadata. According to media studies scholar José van Dijck, our personal data and metadata have become the currency we use in exchange for our information and communication services, including our security.[7] But the processes of accessing these sources of data as evidence are undergoing massive social, political, and institutional shifts that can be seen in the archival logic of storing digital information and the power asserted when controlling access to it. This book tells the history of this upheaval: how for-profit companies gained control of born networked data through collection and data management with a series of archiving machines and distancing techniques, disempowering individuals and public institutions from accessing information they create.

Telling this story is important not just because it gives us a richer history of our information society and the impacts of datafication on the practices we engage when interacting with information and communication technologies (ICTs). It also illuminates how we know what we know through contemporary data sources and the data archives growing at the edges of tomorrow.

The scenes of data management interrogated in this book are archival, but not in the traditional sense. Instead of examining institutions with long traditions of preserving knowledge and providing public access to information, I focus on scenes of data management where digital data begins, originates, and comes to be standardized. Unlike nonprofit or public archives and traditional information institutions, I examine the archival influence on data management found in the early digital storage formats and computing innovations of research labs and mobile tech startups. The book shows how many of these innovations have resulted in data archives where much of their assumed value is in their aggregation for reuse and potential for prediction, where the collection of data becomes nearly constant and taken for granted but access is constrained and used to generate profits.

ARCHIVES, ARCHIVE, ARCHIVING

The word "archives" conjures up a range of places, institutions, and social structures that few other words can. What do we mean by archives, archive, and archiving? It is worth providing some periodization and lexical history of these terms used throughout the book to set the stage for studying archives of data.

The earliest premodern and ancient archives were made up of many of the same kinds of everyday records that we keep today: receipts, correspondence, administrative actions, land surveys, laws, and property records. In prehistory, ancient archives held records of taxes and tributes, and stores of grain, even honey.[8] In the Middle Ages, a more selective approach to recordkeeping emerged with cultures of handwritten transcriptions and copying manuscripts. And as medieval literacy increased, a distinction in values can be seen in the active use and circulation of certain types of records. Those used frequently were copied and transcribed by scribes, more often, whereas others were carefully preserved, often under lock and key.

In common usage, "archives" is conceived semantically—that is, as having meaning about a word documenting something that it also refers to. Confusingly, the term is both a form of containment and what it contains. But most importantly, the historic specialist definition of the word is quite at odds with everyday conceptions and the way that it appears in computer systems (such as the option to "archive" an email or a post on Instagram by hiding it from view).

The word "archives" first appears in the English language in the seventeenth century, coming from French and from the Greek verb *archaeon* meaning "to begin, commence, to rule." For citizens of ancient Greece, the archaeon was the official residence of the archon (or arkon), a ruler or magistrate who held custody of public records.[9] For much of the eighteenth and nineteenth centuries, it was rarely used as a singular term, and "archives" came to be understood as both a records repository and the records themselves.[10] By the nineteenth century, the appellation shed its magistral residency connotation as archives began to stand for physical places that stored and protected government records, registries, and chanceries of nation-states.

At the beginning of the twentieth century, definitions of "archive" began to have public qualities ascribed to them as "a place where public records are kept" and "the public's records." An etymological dictionary published in 1924 notes that the singular term "archive" is unique in most languages, except for in Holland, where an archive also serves as a cadastre registry for land ownership of a territory.[11] During the industrial era, when records management became essential to organizational bureaucracy in the administration of both governments and corporate firms, a formal distinction between active recordkeeping for business practices and keeping archives for posterity arose. It is during this time when archives began to be associated with collections of cultural heritage. As archives came to be understood as repositories of knowledge, the evidence they contained needed preservation and protection. Thus, preservation for future access became core archival processes for government, religion, science, pursuing justice, and understanding the world. Archives, as we now understand them, would be housed and managed by formal institutions, and become essential to how we understand their power.

For most of its history in the English language, the term "archives" is used in the plural and primarily understood as a place. I provide this etymological detour to emphasize that for much of its appearance, the use of "archives" (and "archive") denoted places and records, not a set of practices, processes, or actions that we might assume are found in archives today. Even until the late 1990s, lexicographers agreed that "the singular form of the archive is now generally archaic except in names such as the Bittman archive and in reference to particular sources."[12] "Archive" was not described as a common verb in etymology dictionaries until 1999.[13]

Recently, "archiving" has entered our discourse in more popular usage to mean some kind of effort or process of preservation. A common misconception about the word "archiving" today is that it is what archivists do. In fact, the verb "to archive" is *not* commonly used by professional archivists, museum curators, or librarians unless they are managing the transfer of data from digital collections. Instead, "archiving" comes from the middle ages of computing (approximately the 1950s to the early 1980s) when processes of data handling, collection, and accumulation emerged alongside externalized storage media, standardized formats, and processes of exchange and transfer with the rise of time-sharing and other computer utilities.[14]

Today, the *Society of American Archivists Dictionary*, a professional handbook with many editions that is now published as an online resource of term histories and technical practices, explains that the use of "archive" as a verb "means to back up or to store data offline."[15] In professional literature, the term has occurred commonly since the early 2000s. However, the editors note that many professional archivists avoid it because "the verb is sometimes considered a déclassé usage that undermines the serious value of archives and perverts the meaning of what archivists do."[16] There are politics in the polysemy of archive and its use.

Despite its supposed unpopularity in professional parlance and in training manuals, archivists have used "archive" as a verb to describe digital data collections for decades, both the computational processes of storing and handling data and digital preservation approaches more broadly. The earliest examples of archivists using the term "archiving" involve preserving data with magnetic tape and managing tape libraries, especially systems for preserving oral history interviews and field recordings.[17] However, many of these professional publications detailing data archiving systems are directed toward computing professionals, social scientists, and research data managers.[18] This is no coincidence. Archiving in its earliest sense denoted data storage that could be taken offline and stored away from the computing machines used to access magnetic media formats.[19] It is worth emphasizing that the earliest definitions of the gerund "archiving" refer to securing data offline because today archiving in its computing sense is often marshalled by archivists and media scholars in arguments criticizing the impact of both big tech and institutional approaches to preserving digital collections with computing processes that assume constant, consumptive connectivity to the internet and cloud storage services.[20]

Digital media scholar Wolfgang Ernst reminds us that as the term inflates and grows to become a "universal metaphor for every form of storage," it will inevitably drift so that "the term archiving has nothing to do with archiving in its classic sense but rather merely denotes data storage."[21] Because of this drift, the idea of archive can be stated as an axiom: it serves as a container to isolate and create more value in its storage capacity.[22] Many critical scholars have engaged with archives as institutions of memory by emphasizing their jussive power over what can be remembered and forgotten. But much of the power in archiving data is in managing access to it. In this book, I employ a productive inversion by asking how data archiving—those storage and access processes from computing—has transformed memory practices. The forces that draw upon the constant accumulation of data and separating data from originating contexts and creators mean everyone and everything can be an "archive," but few control how these archives are accessed. I contend that the factors that lead to the denominalization of "archive" that we know now are due to grammars of action that shape the means of interacting with and conceiving access to data once it is created.

GRAMMARS OF ACTION

Looking for a "save" button is what Philip Agre called a "grammar of action." A grammar is embedded in a system such as a word processor that conditions people to use computers and systems in certain ways.[23] When my software program acknowledges that I am looking for a missing "save" button, it reveals that I am not yet habituated to the new grammar it seeks to engender. Early generations of computer users expected to insert a floppy disk into a disk drive and either open the file or manually save it to access its contents later. With time, there will be no need for a system dialog to remind users like me about the automatic save process running in the background; it will simply be a reasonable expectation of writing using laptops. But grammars of action impact how a user feels control over (or disempowered by) an information system and whether they understand or trust it. Such grammars can be useful heuristics for unpacking how practices change with the adoption of new information technologies, not just for individuals but also for large groups of people using systems together.

Grammars of action may also nest in, subsume, or mimic prior action from legacy systems by incorporating routines. For example, prior "save" buttons from PC software featured icons of floppy disks, mimicking saving to a disk without necessarily needing a disk drive or a disk to save information. Over time, grammars may maintain artificial similarities to former sets of action, engendering tacit assumptions about processes while creating huge technical or material differences. So, although the action or process may jump, grammars remain encoded in deep processes of technical activities. In the Macintosh OSX operating system, the Ctrl + S key combination saves a file to memory, but Ctrl + S in DOS "stops" and resumes an operation, demonstrating a discontinuity between two kinds of saving actions: one stores a file and the other pauses to save a place in a process.

With this book, I show how attending to such grammars of action embedded in our information systems reveals discursive regimes of access to data once it is represented, stored, and "archived." By tracing grammars through vignettes of data representation and storage, particular formats allow us to periodize genealogies of information technologies such as operating systems and media formats to store and manage data. In locating and unpacking actions tied to data storage interactions, we can more fully account for accessibility, its design, ethics, and a way of knowing data as it is preserved. As I contend, the means of storing data impacts the possibilities for managing access and asserting control.

This book shows how grammars of action for saving and storing digital information can be found in what I call "archiving machines," which instantiate powerful techniques of distancing. Machines are technical apparatuses that are used to support tasks within structured processes, such as sorting boxes or assembling car parts. Archiving machines are apparatuses connected through layers of networks to support the saving and storage of information, ultimately preserving evidence, that then encode how it can be accessed. Access is significant because archiving machines, like my word processing app or my mobile operating system on my smartphone, contain grammars of action that result in data being archived without a user's awareness, denying some routes for future access while entrenching others. This establishes distance—space between the data people generate and the cloud computing architectures that serve as utilities to store and manage data. Where data is transmitted, stored in cloud storage cases, the grammar of action for automatic saving conditions users to cede control over data access while distancing them from where their data lives.

If some archiving machines produce distances that create more space between people and data, other distancing techniques may bring users closer to encounter machines with interactive features that engage conscious, even tedious (and sometimes contentious) data management processes. For example, when I teach my students about digital preservation, an early step is learning how to engage a write blocker for an object we want to preserve so that the digital information doesn't change when we access it. Write blockers prevent disks or software from being overwritten as they are accessed. For floppy disks, the write blocker is a small plastic tab that can be physically switched and snapped in place, but for software write blockers, a series of permissions and software clients must be engaged with clicks and menu options using a user interface. Both write blockers present interactive features that decrease the distance between people and data being accessed, but these distances are enacted differently because of the way data is structured and formatted. Many students ask why these processes aren't already automated and assume it should be so. But this speaks more to the grammars of automation actively established today. Indeed, many active distances can be seen through the lens of automation or assumptions about the embeddedness of machine-actionable approaches to data management.

The rise of automated and frictionless services is based on the assumption of automatic data archiving. Today, the ability to automatically store and preserve data results in archiving machines controlling access points and other gatekeeping functionalities. A third kind of distancing technique pulls actors together by closing the gaps between data collections and third parties, making it easier for third parties such as data brokers to access data archives through integration and extraction tools. For example, the email app on my phone relies on time-sharing functionalities of platform services and cloud storage architecture. These grammars of action push and pull data services and then not only condition users to expect automatic storage but also teach them to forget how to oppose the politics of asymmetrical access that platforms tend to enforce. For most platform services, a condition of use involves the dispossession of data from users, where services are exchanged for data or contextual metadata that can be stored as data archives and made available to third parties. By bringing industry actors and brokers closer to data collections, this kind of distancing technique builds on the others mentioned above to create asymmetric access between stakeholders by pushing creators away from managing their data. Bringing some actors closer to data archives via access points such as application programming interfaces (APIs) while establishing hurdles for others is a form of power. Asymmetric access thus highlights how firms who play an infrastructural role in data archives have become today's archons—those that govern and assert power through the functional sovereignty of data archives.

Scholars such as Kean Birch, Catherine D'Ignazio, and Rob Kitchin, among others, have argued that the rise of datafication enlarges state power through quantification, statistical reasoning, and computational techniques in the sciences. Yet, the state and science no longer predominantly collect and analyze data using their own storage infrastructures. Much of it is contracted to hosts that manage data. Corporate big tech firms such as Microsoft, Meta, Google, and Amazon collect data, host it, and offer data management services, thereby entrenching their data archiving processes as de facto standards for cloud computing. Many of these data firms integrate and reserve rights to train models and reuse these transformations. Access to data has thus become monetized, incentivized, and transactional as part of the market logics of data-driven capitalism.[24] Yet, in addition to persistent storage, each of these cloud storage products provides tiers of access tools and software integration, such as image recognition libraries and machine learning toolkits. These high technology firms have also become providers of enterprise and cloud data storage to governments and learning institutions in the public sector.[25] The corporate goals and assetization of data in big tech stand in stark contrast to building the state or knowing the world empirically through collections of data and the generation of statistics.

To describe the rise in power of data-driven companies, I draw upon what Frank Pasquale calls "functional sovereignty"—when high technology corporations exercise juridical power through their information services to create a digital political economy.[26] Using cloud data storage is contingent on allowing these companies to collect and access our data to preserve it, further entrenching their functional sovereignty through our information infrastructures. For this reason, data archives cannot be simply thought of as traditional archives "gone digital" and turned into datasets. Rather, capacious data archiving techniques from platforms and corporations benefit from network effects, creating circuits of power that control who can access these caches and how. More simply, the innovation and implementation of data archiving relies on distancing techniques that ensure asymmetric access. Most grammars of action habituate users to accept the dispossession of their data in exchange for automatic data management and, in time, forget that their lived reality is governed by functional sovereignty from the data archives they create that are managed by firms.

CONCEIVING ACCESS

A second heuristic offering I develop throughout the book relates to how meanings of access change with data in different forms and scales. Access is a value that materializes within all information technologies, influencing social practices such as reading, education, knowledge, privacy, and remembering.[27] In each body chapter of the book, I examine how data access has been conceived by different groups of people, sometimes in direct conflict with each other, as they seek to create, manage, protect, and access data. From bureaucrats and government researchers to data managers and graduate students to software developers and scientists, we can see how data born in networks has shaped what we expect from storage and access technologies. I use vignettes of confrontation surrounding data access as an analytic frame for understanding and interpreting the significance of data archiving today.

"Access" enables and limits the power of archives managed by communities, corporations, and institutions. For centuries, archival access involved managing objects and representing documents based on context, existing relationships, structure, order, and form. The power of older archives was constrained by their place and the information they retained, with records often kept secret. Scholars in information studies have argued that access to information has driven public engagement, accountability, and governance. Nearly 150 years after public information institutions began to enact public access in their collections, advances in technology, organizational change, and political landscapes still influence the shifting realities of what is meant by "access."

This book shows how the evolution of archiving and managing data has changed how we conceptualize access too. Data can be formatted, transmitted, redistributed, and streamed in ways that are slippery and hard to control. Born networked data can be accessed in so many ways that leaks and hacks are anticipated, even routine. But as I will show, although data archives today still achieve value through managing access, they do so through mutability, the promise of future growth, and the separation of data and contexts from whence they began.

Access is the animating force of archives. It defines what can be known about, in, and through the evidence within archives as repositories of knowledge. However, commercial data archives now have upset the power of institutions over our digital cultural memory. This distancing has led to accumulation and differential access to data, driven by the inaugural act of data archives over what can and cannot be remembered. I refer to asymmetric access to data that has empowered corporate data archives as "distancing techniques"—an abstract and material concept with ontological and infrastructural dimensions.

Returning to the current dominance of commercial data brokers, they are allowed access to data collected by corporations, insurance companies, government surveillance bodies, political campaigns, and the entertainment industry. Curiously, these oceans of data are not easily accessible to the people who created them. Instead, users are engaged in surface-level, software-driven experiences that extract data and are subject to further enclosure. Metadata, or the semantic data ontologies used to represent data in systems in preparation, are essential for their assembly, organization, and normalization. These ontologies, enclosed within platforms, are collections of user data that corporations monetize by brokering access to third parties, such as technology developers and government surveillance agencies. Consequently, as data is compiled and brokered for access, the grammars of action distance the creators of data from its original contexts. Data from the use of platforms assumes its value by way of such distancing techniques.

The distance from the data we create is assumed to be an inherent cost of today's information infrastructure and networked ICTs. It is largely regarded as a trade-off for the convenience of the cloud and the connectivity that ubiquitous computing supports. This fait accompli perspective denies people power to manage data, since it accepts the shift in access from institutional archives to corporate data archiving. However, we don't necessarily need to be distanced or ignorant of ways to access it. To understand how, we should return to how techniques of distancing emerged at distinct moments in computing history as data management and processing developed. Changes in storage media, increases in collection, and development of new processes of circulation have changed the nature of data accumulation and access and accelerated the public embrace of platform services in exchange for personal information. For some access points, the distance between data collections and the ability to access it is close together, easily negotiated, and well routed. These access points are typically designed for profit-driven data brokers. But for other access points, the distance is capacious and varied, containing obstacles and moats. Understanding how these distancing techniques are implemented and maintained helps us understand different scales of transmission, storage, and access points embedded in data archiving processes today.

Asymmetric access to data has played a vital technological role in today's information society. However, descriptions of the development of these technical and archival processes are largely absent from studies of information technology, computing history, and digital infrastructures. Instead, scholars typically focus on topics such as privacy rights, standardization, corporate histories, and organizational development. Genealogies of data from the state and the sciences tell us about information infrastructures, cyberinfrastructure, and knowledge making with the internet. However, few of these accounts confront the materiality of all this "heavy" data that must be stored and processed in archives and what is required to make meaning and action once it has been automatically saved with archival aspirations. Most of these heavy data stores have been preserved in a small number of places with infrastructure that is increasingly privatized and governed by big tech corporations. The guiding principles of these data centers contrast with policies and preservation mandates of institutional archives, where the precepts of public interest, cultural memory, and archival theory have guided access for decades, even centuries.[28]

Toward the end of the book, I show how data-driven corporations will go to great lengths to assert functional sovereignty and the jussive power of traditional archives when they promise effective data storage and access. However, public values are often at odds with the values inscribed in the architectures of these dominant information technology firms. Many critical scholars who engage with archives as institutions emphasize their jussive power over what can be remembered and forgotten. Applying their scholarship to critique data-driven corporations, this book asks how data archiving—storage and access processes developed in computing—has transformed digital cultural memory.

Some have claimed that everyone and everything can be an "archive," but because of distancing techniques and market logics of private corporations, I argue few can access them.[29] How did corporate data archives, rather than traditional public archives, emerge and evolve by controlling access? What are the implications of their power to control what we know and what can be known from archiving machines? The answers to these questions can be found in the grammars of action from archiving machines such as cell phone apps, shared cloud drives, cell towers, internet company data centers, and earlier networked machines that accessed information on magnetic data tape. The following chapters address these questions by presenting vignettes of confrontation that have unfolded since the 1960s, during which data archiving processes came to replace earlier methods of storage and information management with automated recordkeeping.

METHODS, SOURCES, AND ORGANIZATIONAL STRUCTURE

This book builds on my past research on the interaction of information institutions, computing cultures, and the preservation of digital cultural memory. My interpretive methods use primary sources, computing artifacts, and oral history interviews. Where possible, I draw upon primary sources from research archives, libraries, and museums of technology. I also make use of technical manuals, developers' documentation and forums, instruction handbooks, and published business histories, such as corporate technical documents published by firms such as the Digital Equipment Corporation, Palm Inc., and Apple Computers. My inquiry is especially indebted to computing histories and popular accounts of minicomputers, early mobile devices, and personal computers in the US.[30] Many of the primary documents I use for research have been digitized and are accessible in web archives made available online by the Internet Archive and the Computer History Museum.

The archiving machines and data management processes featured in this book chart a computing history in the United States from 1960 to the present. My line of investigation also coincides with significant legislative events in the US, including the National Data Bank proposal during the Johnson administration, the 2001 USA PATRIOT Act and the subsequent USA FREEDOM Act that resulted in the deletion of the call detail records data archive in 2018 acquired by the National Security Agency (NSA) from cell-phone service providers.[31] There are many ways in which the history of digital data management could be told. Countries such as Brazil, Germany, Denmark, and India have enacted federal data policies to protect the access and exchange of their citizens' data on platforms founded and operating from the US. The accounts presented here engage with some of the international impacts of the privatization of data collection and expanding international markets. It is my hope that this book lays a foundation for research that examines the infrastructural role of data brokering platforms in countries around the world, particularly as the next billion users begin to access the internet with mobile phones, wireless networks, and attendant digital services.[32]

Archiving Machines traces how managing storage and access to data contributes to the datafication of our society. In it, I explore the idea that data archiving is giving rise to new power formations and inequalities that result from asymmetric models of access to the accumulation of massive collections of born networked data. Two sides of the same coin, I show how archiving came to be commonplace, while asymmetric access now shapes our expectations of preserving digital cultural memory. I also theorize what this access regime and its data management realities mean for accessing future archives. The story starts with the reality that today's networked data storage originated in early time-sharing models and virtualization technologies from the 1960s to the early 2010s.

Instead of telling a story of how data archives came to be administered in state or information institutions, this book introduces how public access failures opened space for corporations to seize control.[33] Through vignettes, I show how data archiving came to be the purview of corporations, cleaving access to data from democratic institutions that produced knowledge intended for public access and archival preservation. The story of these institutional failures and corporate successes are mapped across vignettes of confrontation around data's materialities and grammars of action. These explain how new archons of memory have emerged and why they continue to dominate access regimes in the data landscape. To illustrate how powerful distancing techniques provide access to data as well as inhibit information and memory, I locate and present these vignettes across three parts of the book, in which archiving machines' materiality, functional sovereignty, and power were supplanted with the data access regime so familiar to us today.

PART I: CENTRALIZING DATA

To understand the power of archiving data, it is necessary to trace how data became externalized from early computers to standardized storage formats. Part I (chapters 2 and 3) of the book explores the prehistory of time-sharing, centralized data banks, and the 1965 proposal for the US National Data Center. I begin by tracing the data archiving processes resulting from magnetic tape and time-sharing computer models from the 1960s to the establishment of data banks and the rise of database processing and managed data silos in the 1970s. Before data could be centralized in data banks or databases, it needed to be externalized from computing machines with magnetic media that was standard and easily moved.

Once data could be externalized in a standard format, it could become mobile, mutable, and distributed across networks. Chapter 2 focuses on the challenges of storing data with core memory and how the advent of external magnetic tape created artificial memory and external storage. As an artificial storage medium that separates data and file structures from the core memory of the computer, we begin to see the emergence of hardware customization and software development. The chapter begins by introducing the microcomputing architecture, which allowed for time-sharing and many users at once. I then introduce magnetic line tape that created new ways of reading and accessing core memory of computers. Unlike earlier forms of artificial storage such as punch cards, magnetic tapes couldn't be read with the human eye. These digital inscriptions contributed to a first step of externalizing data from computing machines, creating distance while bringing users close to interactive computing experiences.

Throughout the book, vignettes about accessing data collections illustrate how replacing historic archival processes for analog information welded with data management techniques brought about shifts in power. These moments allowed high technology corporations to wrench power from institutions by asserting the functional sovereignty of archiving machines through data collection at increasing scales of storage. But data archives and archiving were not simply transposed from modern archival science to corporate contexts. As mentioned above, archives have loomed large in Western epistemic cultures of science and in the building of bureaucracies. In the US, data generated from postwar federal programs led to proposals for a centralized data bank of government statistics from different agencies in the early 1960s. The proposed National Data Center would be responsible for collecting, organizing, and providing access to public data increasingly requested by social scientists, economists, and other academic researchers. But public fears about how centralized data collected from citizens would be misused by the government quashed the proposal.

In chapter 3, I examine why the proposal for a federal data center failed, paving the way for commercial archiving and data management we see in platforms today. I explore how the public's understandings of personal information, data, and privacy conflicted with the access gap that a National Data Center sought to close by drawing disparate data silos together. I bring our attention to this moment when the state failed to create a national data archive because it created space for big tech firms such as Amazon, Google, and Meta to assert functional sovereignty in and through their accumulation of data and brokering access to data archives.

PART II: SHARING, SYNCING, AND STORING PERSONAL DATA

Central to arguments about contemporary data privacy is the belief that users should have control of their data, or at least more control over how it is used by third parties. But total control of one's personal data is an exception. Part I of the book sheds light on the relatively brief period during which users actually did have control of their data—roughly, the 1970s–1990s. This background is crucial for addressing normative assumptions about who should have control of digital data, personal or otherwise. The government's forced retreat from centralized data control created a vacuum that was filled by commercial databases, which is important background to understanding the realities of contemporary data management driven by wireless networks.

The next section of the book concerns how archiving processes related to information storage and file structures paved the way for the data we create as individuals today to be both digital and born networked. I begin by introducing grammars of action found in personal computers introduced in the 1980s that led to the rise of the personal digital assistants (PDAs) and, eventually, cell phones in the 1990s.

Focusing on the smallest units of information created and stored on personal devices such as text snippets and address book entries, chapter 4 examines the transition from files to apps. As digital data moved to a format that was bundled and structured as files, it eventually became accessed in programs on personal computers and apps on mobile devices. Where earlier eras of information ownership, privacy, and access relied on an individual to take charge of their own "personal information management" (PIM) with structured documents and file structures, apps would transform who owned what.

Internet access and, eventually, mobile networks subsumed the responsibility for tacitly collecting personal data traces in exchange for supplanting control over personal recordkeeping to externalized platform and cloud services. By exploring this transition from personal data management to personal information management applications supported by corporate actors, we can see how the political economy of data archives was altered in the move from individual recordkeeping to data services. Part II shows how storage structures in pre-internet early personal computing devices conditioned users to see data as a gift exchanged with computing firms. This grammar of action already distanced users from their personal data by encouraging them to see it as something to exchange for software services.

PART III: ACCESSING DATA TODAY

The third section of the book describes the rise of platforms and their influence on our ability to access data. Their rise illustrates how corporations have seized control of access through archiving processes found in virtual containers in cloud-based information services and enterprise data centers over the past twenty years. Chapter 5 examines another failure of government data archives for surveillance and the rapacious data gathering that occurred via telecommunication networks: the NSA's call detail records program that was ended in 2018 after it was discovered that this "haystack of haystacks" had gathered too much data to be useful in detecting and preventing domestic terrorism. In 2013, Edward Snowden leaked documents regarding surveillance programs that were built on top of the collection of call detail records (or "mobile telephony" records) of American citizens from mobile service providers such as Verizon and AT&T. Later, this data collection program was found to be useless, not because of its scope and collection methods but rather because once reassembled away from its original context, this intelligence data lacked meaningful evidence. Instead of dwelling on the legal status of the data collection, I focus on why one of the largest known surveillance datasets of communication activities failed to result in actionable evidence for the NSA: because the archive was "too big to know."

In chapter 6, I turn to the rise of platforms as new archons that actively manage data collections they have extracted from users. Increasingly, platforms such as Meta, Microsoft, and Google present themselves as access points, providing scientists, researchers, and journalists with data access to spur technological innovation, carry out scientific research, and understand news and current events. I show how the archival logics of the democratization of access and bureaucratic management are increasingly exercised by corporate firms through data exchange, interoperability, and transfer through APIs and data brokering, as well as through cloud storage and tacit personal data collection for data integration. I show how the grammars of action from mid-century computer processing, data management and centralization, and personal information management from the early 2000s have engendered constant data collection without meaningful access for individual users creating knock-on effects for researchers and institutions' seeking access to platform data for scientific inquiry. The result of failures to create public access points explains why data archiving processes are now shaped and controlled by a few sets of players who profit from asymmetrical access to data collections.

AN INVITATION

My intended readers for this book are concerned with how we will be remembered by future peoples and machines. As an information scholar concerned with archives and preservation practices, I am fascinated by how we access recorded information—when we use our ability to read to access information, when we rely on expert systems for access to facts, when we trust stewards to access knowledge, and when we rely on machines for access. Traditional preservation mandates found in knowledge institutions employ modern archival theory that originated in the early twentieth century.[34] They are largely concerned with how recorded evidence from the state, knowledge institutions, and epistemic cultures are circulated and united with recorded information to preserve knowledge for access.[35] But these historic mandates and expectations are being turned inside out with data made from networks. And the fastest growing data archives are now found in high technology firms, cloud storage services, and data-driven corporations. Firms provide or inhibit access to data to grow revenue in ways that break apart institutional norms of information ownership, responsible stewardship, and public access to knowledge. This disruption is evident in the wave of new artificial intelligence (AI) technologies, which depends critically on access to vast amounts of training data. What do these changes mean for the social, cultural, and institutional processes that rely on earlier regimes of access? In this book, I argue that the consequences of this change can be seen in how information is now automatically saved and archived as data, and who controls access to it.

Archiving Machines is not about traditional archives found in institutions but rather about how accessing knowledge has transformed as computational processes for storing and managing data developed, thus putting pressure on concepts of archives, archive, and archiving. In other words, this book explores how "archive" became a verb with the rise of data-driven recordkeeping to unpack the stakes of archiving machines. The increasing role of digital data in the production of knowledge is well known to scholars of science, scholarly communication researchers, and computing and information sciences.[36] Automated recordkeeping changed management and privacy rights. But there is no history of it from the perspective of active recordkeeping with data. And as I show, recordkeeping with data technologies is, just as it was in antiquity, about keeping archives. But the value and place of accessing archives of data have been altered.[37] In this sense, a book on data archives serves as a corrective to normative visions of what an archive is and what archiving means now to cultures of data. The implications can be seen in how information now automatically stored as data in private storage infrastructures and how access is governed through the functional sovereignty of big tech.

Through each of these confrontations, I build a theoretical understanding of the preservation and access of data by examining epistemological shifts in people's understandings of information and evidence found in late twentieth-century computing technologies. I aim to show how archiving machines link the politics of data to both grammars of action and access regimes. In particular, the largest tech firms and platforms, especially those leading in AI innovation, are poised to benefit the most from controlling access to archives of data. By following these data archives and archiving processes, this book spotlights technoscientific encounters where memory, future, storage, and circulation have unfolded where grammars of action originate, get taken up, trusted by users, and managed by firms. These sites, I contend, are crucial to our understanding of data access and, ultimately, who controls what we can know.

2

DATA PUNCH: FROM MANUAL TO MACHINE

Often, when we think of "data" that is collected and captured, we assume it is digital and handled by machines.[1] But until quite recently, data was made manually and relied on intensive physical processes. It was collected, counted, structured, styled, sorted, and stored by hand. For much of history, data took form in organic materials by way of human gestures and actions. Using rock walls, leather, papyrus, sticks, bones, and clay, we inscribed data *on* material forms to become information.[2] I purposefully write "on" because when writing today, we may we think of applying ink or pencil graphite *to* paper. Similarly, most data creation and information storage in prehistory involved brute force from impressions, cuts, scratches, and dimples made on the surfaces of organic materials.

The history of data is part of a much longer history of information systems, knowledge production, storage techniques, memory practices, and institutional governance that has relied on records that were manually assembled over decades and even centuries. Records—the containers that data are collected and stored in—reveal a society's relationship to state power and how individuals are represented by institutions. For most of time, recorded knowledge has been produced by relying on manual data practices and storage formats. Indeed, it is fair to say that from cuneiform tablets to tallying sticks, data practices have involved handling objects and manually making impressions on their surfaces to record information. I begin with this because data was incredibly tactile, all the way to the punch.

The punch is a way of inscribing and storing data in cards that allowed records to be processed electromechanically by machines, resulting in new data resources.[3] Data created from a punch mechanism and stored in collections of stacked cards allowed for information to be counted, sorted, combined, and processed using mechanical tabulating machines at previously impossible scales and speeds. After the punch, data practices would still rely on hands, but the new manual processes of punch cards largely served to make data accessible to machines.[4]

The history of data management in the United States often focuses on the development of database management systems and new statistical methods that faster processing and assembly afforded by digital data storage. This transition to digital storage formats was when archiving became a verb, partly because of new storage technologies for large datasets combined with the appearance of machines that could handle and process more and more data because of these formats. For long-term preservation, it is these early moments of origination and stabilization of new information that shape the potential for future data access. Tracing the transition from one information age to another reveals whether information objects will be preserved for intentional future access. These moments of transition between formats reveal why successive access to information objects becomes codified into power structures in systems.

This chapter unearths a pivotal moment in twentieth-century data management that relies on how data is stored and formatted: the transition from manual data transcription with punch cards to machine-readable data and automated recordkeeping systems.[5] But moving from cards to magnetic storage is a long transition because cards were to hold data alongside analog paper files, data tape, magnetic drums, and other storage formats for nearly a century before they were fully phased out for digital versions. To shed light on the transition from the mechanized handling of analog information to preserving digital data, I tell the story of National Data Center proposals in the US and their failure during the 1960s.[6] These proposals and subsequent controversies over federal and state data banks reveal how archiving data became a part of government agencies' recordkeeping administration and collecting information.

The debate around a nationalized center for government information collected from and about citizens also marks the moment when archiving data for access became a concern for scientists, bureaucrats, and the American public. At stake was whether personal information gathered from citizens by the state could be processed and transformed into aggregated data resources for access without violating Americans' privacy rights. Critics argued against combining, connecting, or migrating records to data banks for fear of bureaucrats abusing access once computers, automated recordkeeping systems, and databases full of Americans' personal information could be linked. This moment came far earlier than the data protection, data access, and privacy laws proposed today, but the debate plays a significant role in how data is both collected and accessed from individuals by corporations in the present. As such, it tells us much about how the public grew aware of the government's recordkeeping practices with little scrutiny toward corporate data collection and archiving efforts, and the power in massive amounts of personal information collected that can be aggregated with computerized recordkeeping systems.

Far before data banks and machine-readable records entered the public consciousness, there were privacy threats and concerns about government use of computers. The proposal arrived during a frothy moment for information policy and the history of privacy legislation in American history. Previous scholarly research on the public outcry and congressional attention to the National Data Center has largely focused on privacy policy and its role in information privacy (ironically creating more records rather than fewer). Most privacy scholars focused on this era, such as Sarah Igo, Priscilla Regan, and Michael Schudson, tend to mention these National Data Center proposals within the history of the "right to know," government transparency, or the "right to privacy." These privacy scholars attribute the public scrutiny with the centralization of government data to increased government transparency outcomes. At the time, courts were expanding people's information protection and privacy rights under the First Amendment, with the FOIA of 1966 and the Privacy Act of 1974, respectively. Alongside this scholarship on legislating privacy and transparency, I argue that mid-century data management can be connected with US archives and libraries in universities and with information processing and service centers for social science research that are often left out of these accounts.

In this chapter, I argue that the National Data Center failed because it lost a war over the public perception of personal information stored in files—who collected them, how they were aggregated, and, once computerized and stored in digital formats, who could access them as data managed in archives. By tracing these early visions of access to files and punch card data, we can see the power in material forms that contain and bear information.[7] For decades, government agencies used data processing machines to tabulate punch cards, but once machines were put to service for archiving digital data with information aggregated from individual citizens, conflicting visions about access took the spotlight. The public awareness of this transition and pushback to the proposals, in historian Sarah Igo's words, "revealed a transformation in Americans' stance toward their information and powerful authorities alike."[8] The failure of the National Data Center captures a conflict about storage containers and their purpose enabling access to evidence, revealing how data, once stored in interchangeable formats such as punch cards, can be made meaningful by both people and machines.

PAPER DATA AND PUNCH CARD COMPUTING

Since the 1930s, data banks of punch cards had been used for tabulation and analysis, but these collections largely supported rote calculations such as counting votes or financial transactions. Mainframe computing architecture at the time made it incredibly difficult to quickly query large datasets. Such dynamic calculations were essential to making predictions, especially of near-term events such as the weather. Data was processed in bulk, first punched manually and then verified by operators at machines. The largest data collections were also the most difficult to work with because they needed more computer time to process thousands of cards that could jam, tear, and fall out of order. But as computers became faster and more efficient at processing data at scale, recordkeeping migrated to massive sets of punch cards processed by machines. Punch card computing opened new possibilities for any sector that relied on collecting and aggregating information for operations, from scientific inquiry to government administration and business management.

At the time, the terms "data archive" and "data bank" were used synonymously to refer to collections of paper files, punch cards, and eventually both paper tape and magnetic tape. Whether a data archive or a data bank, both terms were frequently applied to different organizational arrangements but generally referred to an active repository or service center that provided researchers access to collected data. Importantly, data banks could be used in business organizations for recordkeeping. In time, services would expand to generating summative reports for managers. Although the data stored in most US data banks was considered "manual," it wasn't always machine readable until the 1970s. Instead, datasets and studies were synthesized into literature such as reports and topic abstracts by data librarians—published resources for researchers and administrators.

In the 1940s and 1950s, social scientists and government bureaucrats began to focus their collective attention on growing caches of machine-readable data in government, debating how these new collections in data banks could be accessed for research. The largest sources of data were still in the federal government, in part because they were the only organizations that could manage the storage and processing. Throughout the 1900s, the US government had collected a range of structured records and information sources that were aggregated into statistical data. These records included data on tax returns, social security records, census forms, housing data, and birth certificates. Despite the size and scope of these caches, government data banks were the most difficult to access because accumulation was prioritized over access and quick retrieval.

During the punch card era, it was terribly difficult to access data collected across federal agencies and departments, even for the government researchers who worked for and administered them. Despite access hurdles, government researchers developed new statistical techniques for analyzing population data, developing probability sampling and random survey methods as early as 1937.[9] However, federal agencies are separate organizations with different data collection practices all subject to different records laws. Each agency acted as their own kind of records island, with different organizational processes for generating, collecting, storing, and cleaning information according to their own mandates, often governed by legislative policy.

Elected representatives, government administrators, and statistics researchers desired more coordinated access among agencies as well as formal processes and policies for accessing accumulated data. Even so, data banks maintained by the federal agencies were notoriously slow. If coordinating knowledge about the existence of data resources was a challenge, accessing it from separate agencies was even harder. A study of 858 federal agency data banks carried out by the Senate Subcommittee on Constitutional Rights found that requests were evaded because "extracting information was like pulling teeth."[10] To make access issues worse, the data collected by government agencies was frequently redundant, but the information collected was nonstandard. So, it was difficult to manually search data once collections were aggregated and assembled. Most types of requests for aggregated information were simply ignored because of the constraints posed by manual processing and integrating data sources.

MID-CENTURY DATA BANKS

Siloed data banks with poor data access to survey information wasn't limited to states and government agencies. University and marketing research data banks were full of analog data from decades of public surveys that polled opinions of Americans. Survey research methods had been used in the private and public sectors to poll everything from voting trends to toothpaste since the 1940s, and data banks focused on survey research were mushrooming up everywhere.[11] The largest and oldest machine-readable data archive was the Roper Public Opinion Research Center.[12] At its height of punch card computing, estimates report that the Roper Center maintained more than ten million punch cards' worth of surveys and polls—about one gigabyte of digital storage today.[13] Still, running calculations with hundreds of thousands of cards involved manually punching and then verifying them with human operators. It was tedious and incredibly time-consuming work that required teams of machine operators, expert researchers, and engineers to manage the data processing machines.

Despite all the human resources necessary, academic and municipal data banks that existed served local researchers only in limited ways. Since the 1940s, data archives, such as the Roper Center and Michigan's Institute for Political and Social Research, had been collecting surveys, panels, and polls and storing them on punch cards.[14] But accessing data after it was deposited for secondary analysis was challenging for researchers because of the lack of awareness about these resources and standards for searching them. Even local researchers didn't always know what their data banks contained, and scholars beyond these institutions were rarely aware of their existence. The data banks themselves were geographically bound, and users were meant to come and visit them as places. Many early data bank managers lamented that most potential researchers, graduate students, and the public were not aware of these troves of collected information and would likely duplicate data collection efforts without more publicity.[15]

Another serious issue in managing mid-century data banks composed of punch cards was the physical space the cards consumed when stored in shelving and stacks. Hundreds of thousands of stacked punch cards took up considerable space, and their heft stressed the foundations of buildings. Datasets were frequently "weeded" or thrown out to make space for newer collections of punch cards once the data had been tabulated and verified, with the reasoning that published tabulations would be sufficient evidence for future reference. For example, more than eight million punch cards from the 1930 census were destroyed because of space constraints at the National Archives and Record's Administration (NARA)[16] after tabulations of the processed data were published in thirty-three book volumes.[17]

Balancing acquisitions and physical storage of data with transforming collections into accessible information resources became a primary goal for early data managers, librarians, and archivists. Early data banks acted as information processing facilities not just to initiate research also but to facilitate research services on an ongoing basis for the organizations they served. Data banks began bringing in new sources of data to provide access to them as information resources, like libraries and archives. Soon the librarians and research data managers that worked in data banks began to synthesize data and transform it into formalized research, much like other research organizations that would churn out new knowledge in published summaries, abstracts, and reports.[18] Many surveys and poll data collected by data banks were beginning to be transcribed and stored on machine-readable data formats such as tape and microfiche by the 1950s. Unlike mechanized punch card tabulation, which was incredibly time intensive, data stored on magnetic tape reels cut down on processing time, which opened new avenues for secondary analysis and method design, including modeling, prediction, and interpretation.

Tape reels sped up processing data and made data more programmable so that datasets could be winnowed, combined, or prepared for secondary analysis. New sampling methods emerged because researchers found that the frequency of data collection (on tape) could be processed at faster speeds with less wear and tear than punch cards. Computerization decreased the time needed to process and verify data and increased the quality of digital data. But even with new information processing techniques and organizational mandates to initiate and facilitate research, low awareness among researchers and the high costs of storing (and migrating) paper data in collections raised questions about its intended purpose. What were data archives for? Who ought to have access to them, and how should support be provided by research data managers, archivists, and librarians?

DATA MANAGEMENT PROFESSIONALS AND RESEARCH DATA ARCHIVES

Because there were storage challenges and servicing problems associated with managing massive amounts punch card data. Most data banks were eager to migrate collections and address operating resources toward growing tape collections. Ambitious data archives managers proposed ways to improve search techniques for social science researchers, even assembling new datasets and developing research services. For example, instead of waiting for researchers to deposit data, once an archive had acquired robust and reliable collecting areas, the organization could acquire data sources de novo or even purchase data from other research organizations to provide access.[19]

Ralph L. Bisco was a leading visionary of research data archive development. He served as the first director of technical services of the Inter-University Consortium for Political Research and later founded the Council of Social Science Data Archives (CSSDA).[20] Bisco wrote a series of influential publications that surveyed the state of the art among the major institutions who managed data banks, and policy among professional organizations that brokered access to data archives. By the early 1960s, data management professionals began to make distinctions between types of data archives. Some were general-purpose archives, where data would be deposited by researchers and then made available to the complete academic community, like a university research library. A second kind of data archive was known as a "service center," in which the acquisition of data and its later access supported an organization's operations and services for analysis. For service centers, access was typically limited to users based on their geography, institutional affiliation, or security clearance.

Bisco published *Data Bases, Computers, and the Social Sciences* in 1967. The edited volume, a collected proceedings from an early CSSDA meeting of data managers, described how data management professionals could support social scientific inquiry with aggregated data and information processing techniques.[21] Social science data archives were the result of different sources of machine-readable information across different stakeholders and technical developments in computing but primarily involved information that had been aggregated from large populations of people. New information services supporting secondary analyses were also driven by the decreasing costs brought by computers with faster processing capabilities. Bisco predicted the emergence of data banks from commercial marketing and survey research firms (which today we would call "data brokers") and the rise of social research centers in universities and research institutes. The increasing availability of machine-readable information produced and published by government agencies drove demand for accessible research data from data banks.

Despite the growing public- and private-sector interest in assembling data collections for research, accessing information from social science data archives was still a new idea among scholars and scientists.[22] Early proposals from professional organizations such as the American Economic Association discussed conventional approaches but left standardized processes and workflows to be sorted out by data collection agencies. In addition to being stored in cards and tapes, collections would need descriptive information about their context and use from the originating scientists, such as codebooks or the original survey instruments. More cleaning and preparation from data managers and archivists would also be necessary upon deposit to an archive because "few collections come into an archive in anything approaching an ideal form."[23] Just as it is today, data collections acquired by research archives are not immediately ready for reuse. Instead, they often need to be prepared with vetting, authentication, error checking, and metadata creation for both preservation and retrieval.[24] To go from service centers to general-purpose archival collections, social science data archives needed to standardize best practices. Today, these data management practices are well developed and designed by professional data managers and domain-specific research data organizations. But at the time, de facto standards were developing in the transition from manual storage to digital tape as data was collected and then stored for access and reuse.[25]

The focus of development proposals from research organizations, such as the Social Science Research Council (SSRC) and the CSSDA, was on building up repositories, enacting new collections policy for researchers, and predicting use-case scenarios. Proposals emphasized that social science data archives primarily functioned as acquisition sites. They brought in data by working closely with scientists to deposit their data, as much as helping others access it. Once information was acquired, it had to be processed or prepared for its distribution and analysis, which required special plans for storage, maintenance, and long-term preservation. Storage was another cost constraint that was mentioned frequently at the time, in addition to machine processing and the size of data acquired because with thousands of punch cards, "it may cost more to store the derived measures than to reconstruct them each time they are needed."[26]

Although information processing techniques dominated the early professional literature read by data managers, improving information retrieval would be how social science data archives expanded new research services.[27] Ideally, a researcher would request generic data for specific questions or consult with a staff research specialist to design an appropriate query that would leverage the high-quality data collections in the archive. Many of these early data archive proponents would discuss potential issues with designing data retrieval systems. While privacy and confidentiality were mentioned in acquiring data resources and acquisitioning it into collections from scientists depositing data, they would not be a major design consideration in information retrieval or access proposals in early data archive development. Instead, primary topics in handling access typically involved winnowing down aggregated data from large sets into a precise slice or, conversely, braiding together multiple sources. Yet, all these plans from data managers for supporting research services had little practical relevance without a physical site for research access. Much of these professional concerns from researchers, data managers, and data archivists seeded the ground for calls of a nationalized effort at centralizing data and providing access to government information through a federal data bank.

PROPOSALS FOR A US NATIONAL DATA CENTER

In 1962, the SSRC organized a formal study on the use of statistical data for social research in the US. Richard Ruggles, an economist from Yale, was appointed by the SSRC to lead a committee made up of leading social scientists. Together, they would explore the "archival problem" of providing access to existing economic data by focusing on machine-readable information stored in data banks housed in the federal government as well as data collections from universities and research institutions.[28] Accessing machine-readable information found in data banks added a new dimension to data analysis because computational processing had the potential to transform methods of inquiry in understanding society. For social scientists, the archival problem of aggregated data wasn't researcher awareness, storage formats, or information handling processes that primarily concerned data archivists providing access and managing data banks. To them, the main archival problem was accessing data from the US government—the federal government's plans for preservation and coordination to support access that would spill out into the spotlight for most of the next decade.

The advantages of a national centralized repository for data were clear to these scientists: access to more data would be more efficient and valuable. It would be available for more people and from more sources, instead of a patchwork quilt of separate data banks. But before a centralized repository of government data could be developed, the case for a centralized data archive in support of governance and administration needed to be made by bureaucrats to administrative officials and elected representatives in Congress. A series of exploratory case studies were conducted throughout the 1960s to research the problem space of centralizing access to government data. In response, social and behavioral researchers—both within and outside the federal government—proposed what came to be known as the "National Data Center."

From 1962 to 1964, the Ruggles committee studied issues related to preservation and use of data collected by federal government agencies. Their report, *Preservation and Use of Economic Data* (known as the "Ruggles report"), was submitted in 1965 to the Bureau of Budget. Their three-year study found that even as computers were becoming more affordable and prevalent in government data collection operations, there were no systematic best practices for enabling access to machine-readable data once it was processed by agencies. The committee also reported that although the National Archives was beginning to provide access to machine-readable data, it did not have interagency authority or the ability to process such archived data at scale. In the Ruggles committee's view, there was a need for a new "interagency authority" to lead the development of new technical services and administrative arrangements for federal statistics data.[29] Otherwise, we would end up with a perpetual patchwork model that already existed.

Recommendations from the Ruggles report started to gain traction, and public–private partnerships were organized to enable access to statistical data for policy development, government administration, and social science research. The Bureau of Budget agreed that the decentralized information practices of the federal government inhibited its efficiency.[30] For bureaucrats at the large agencies collecting information and generating statistics, the preservation and use of data was a matter not just of access for research and policymaking but also of fiscal responsibility; the preservation and coordination of data archives would save not only time but fiscal resources as well.

Across the twenty-one major agencies in the federal government, the offices that stored the most amount of statistical data for their operations, such as the Census Bureau and the Internal Revenue Service (IRS), accounted for the bulk of the federal budget.[31] Each agency had different handling procedures regarding data collected from citizens, and different access and storage protocols for how long data was to be saved. For example, the Ruggles committee found that at the Census Bureau, vital statistics from County City Data Books (three hundred thousand cards and three tapes) were permanently archived and had unrestricted access, whereas data on farms from the Quinquennial Census of Agriculture (8,900,000 cards) were kept confidential and destroyed after six years.[32] Although larger data collecting agencies, such as the IRS and Census Bureau, had standard policy for handling and restricting access to data, there was still no federal guidance on aggregating data or providing systematic access to collections of data collected by the government.[33]

Upon the receipt of the Ruggles report, the Bureau of Budget contracted Edgar Dunn to form an ad hoc committee consisting of representatives from the largest bureaus collecting data, including the Statistical Standards, Labor Statistics, and the Census Bureau. The Dunn report was submitted to the Bureau of Budget and published in December 1965. Proposals from the report focused on cost-saving investments in computing technology and information services. It recommended file storage and management of machine-readable archival records, referral, and reference resources to describe existing collections, and access services to facilitate data for statistical research.[34] The Dunn report rejected the "naive databank concept" of migrating all data stored on punch cards to magnetic tape. Instead, it focused on how the transition to tape could improve the speed and efficacy of policymaking and government administration, particularly in the future with aggregated data. Although reaffirming the recommendations found in the Ruggles report from the SSRC, the Dunn report for the Bureau of Budget went further and proposed that the National Data Center be established in the Executive Office of the President. The National Data Center would coordinate information flow and data deposits but remain separate from data collecting agencies such as the Census Bureau.

In response to these recommendations from the Dunn report and Bureau of Budget, the Johnson White House appointed a presidential task force to further explore storage of and access to centralized government statistical data. Chaired by the economist Carl Kaysen, the task force held committee hearings in early 1966 and would later publish another report of its own.[35] The Kaysen committee proposed a significant name change. Instead of a "federal data center," the proposed data archive should be known as a "National Data Center" to suggest that both state and local government information would also be included. The Kaysen task force report clearly states that individual privacy would not be at risk because of previous, ongoing federal data collection standards but concluded that further organizational and legal safeguards should be developed once data had been aggregated for federated access.[36]

The Ruggles, Dunn, and Kaysen reports propelled the idea of a National Data Center to support the access and long-term preservation of centralized government statistical data in the US. Each of the committees found and reported the serious limitations of accessing federal data in the 1960s by studying existing processes and policies, in addition to interviewing key stakeholders. All three reports agreed that public access to government data was typically slow, arduous, and filled with gaps. Even when it was obtained, data was often locked into analog formats that were difficult to duplicate and manipulate. The Ruggles report captured the significance of the decentralized nature of federal statistics and emphasized the need for coordinating among agencies for records management and data integration. While the Dunn and Kaysen reports emphasized the need for standards development and technical services from the government outlining data access. Standards and access services would prevent agencies from duplicating effort and increase efficiency in data collection and management, particularly with new storage formats such as magnetic data drums and tape reels. The three reports suggested the government do what research data managers and social scientists had been clamoring for nearly a decade.

According to all three expert committees, the proposed National Data Center should primarily concern itself with developing and coordinating information services that would provide access to machine-readable formats generated by all government agencies. Each of the proposals focused on how a nationalized effort at data archiving could provide guidance, standards, and staff support on how to preserve, organize, and eventually duplicate data so that bureaucrats, policymakers, and government researchers could have easy access to collated, authentic, and up-to-date federal and state-wide data. The Kaysen report took particular care to discuss disclosure standards among data collection and ongoing government regulation that prohibits accessing individuals' information, as well as how confidentiality was ensured through processes of tabulation and aggregation. In its conclusion, the Kaysen report discussed privacy rights and ensured the confidentiality of individuals represented in data that would be stored in the National Data Center. Specifically, the report mentioned concerns regarding information from individual "dossiers" from the Federal Bureau of Investigation (FBI), military service records, government personnel records, and police reports.[37]

The dossier had begun to take on a new significance in the American public's imagination. At the time, there were new social welfare programs and legislative mandates that would benefit from a centralized data archive of federal information. Many of President Johnson's "Great Society" programs—such as the Civil Rights Act of 1964 and the Voting Rights Act of 1965—all necessitated accurate and timely data to manage and enforce. A federalized data repository would increase their impact, but centralizing government information from files collected about citizens could also open the door for partisan critique and public scrutiny.

The data center would not collect any new information that was not already used by federal agencies.[38] However, in proposing to aggregate and centralize it in a massive archive for coordinated access, all three proposals placed a spotlight on all the different kinds of personal information that the government regularly collected about Americans that could be potentially stitched together to create new forms of aggregated information. By claiming that privacy and confidentiality safeguards were accomplished at the point of information collection, proponents of the National Data Center proposals made assumptions about where risks to individuals' privacy would arise. None of the proposals considered the threats to privacy after data had been aggregated for access and reuse.

COMPUTERS, DATA BANKS, AND THREATS TO PRIVACY

During the postwar period, more personal information was being collected about Americans by institutions than ever before, including school records, military service records, and personnel files. Even without machine-readable data and computers, federal recordkeeping practice appeared to invade citizens' privacy by collecting detailed personal information that wasn't ever used. In 1967, a Senate Subcommittee on Administrative Practice and Procedure surveyed all federal agencies about the information they collected about individual citizens resulting in agency dossiers. They found the federal government had collected more than three billion records for nearly every living American—as well as many long dead. The committee concluded that "the majority of Government forms require either nonessential or too detailed information from the individual citizen."[39] In the postwar period, the public was just waking up to the possible dangers of the government collecting information from individuals and making it machine readable. According to historian Jill Lepore, "the debate over the National Data Center marked the first time in American history that the public had engaged in a sustained conversation about the consequences of the aggregation of personal data on computers."[40]

Even though the proposals for a national center did not recommend collecting new sources of data, the power of personal information once aggregated in to machine-readable data had begun to loom large in the consciousness of the American public, who were just beginning to learn about computers in industrial and organizational applications.[41] By the early 1960s, the IRS, the FBI, the military, and colleges throughout the country used mechanized automated recordkeeping systems, giving data aggregation the patina of government secrecy and administrative control. So, as recordkeeping systems in credit bureaus and high schools became computerized, the privacy of citizens' information became political. For most Americans, the new capabilities of computers and public descriptions of them conjured secret dossiers and fears of Big Brother. Aggregated computerized records in data banks became, in Igo's words, "the privacy-invading technology that would surpass all others in Americans' imaginations was the data bank."[42]

Following these National Data Center proposals to provide access to statistical data from the government, several congressional hearings in the House[43] and Senate[44] were held throughout 1966. Lead by Senator Edward V. Long and Congressman Cornelius Gallagher, respectively, the hearings addressed the possible invasion of privacy that would result from a data center using computer technology and automated recordkeeping to manage data gathered from the public. In the beginning, the hearings primarily concerned the impact that computational processing and automated recordkeeping would have on the individual privacy rights of American citizens. But like earlier hearings on the invasion of privacy and automated recordkeeping lead by Congressman Gallagher, the committee "evoked a series of images that would play on popular ambivalence toward computers."[45]

Instead of focusing on the information handling processes and data aggregation techniques that were being proposed to provide more efficient access for existing data collections, the hearings began to air assumptions about how computerized systems could be used to harm individuals by misusing files with personal information increasingly stored in machine-readable formats. As the hearings continued and critiques in the press began to circulate, concerns shifted further from data handling to automated processing. They moved from focusing on the potential impact of new computing technology on data processing to the sheer volume of information being collected about individuals, particularly regarding their "dossiers" and how government data collection of personal information should be curbed.

According to the privacy scholar Priscilla Regan, "Congress's first discussions concerning computerized record systems cast the issue in terms of the idea that individual privacy was threatened by technological change."[46] But soon, the congressional hearings and public debate would shift to the notion that privacy was threatened by aggregating digital data into machine-readable formats. By the end of the year, the congressional inquiries exploded into a full-blown controversy, and for observers, the plan "acquired the image of a design to establish a gargantuan centralized national data center calculated to bring Orwell's 1984 at least as close as 1970."[47] The invasion of privacy debate played out in the congressional hearings and coverage in the news. In the news media, the public outcry was focused on the significance of accessing federal data collected from citizens and the threats (both real and exaggerated) to American civil liberties in the face of automated recordkeeping. These fears about files aggregated to dossiers and accessible through networked computing resources would shape data protections in the US for decades to come.

On one side of the debate were boosters of a centralized national archive for data that prioritized access and standard processes. For years, government economists and statisticians, academic researchers, librarians, data professionals, and federal bureaucrats promoted the significance of having centralized access to reliable, authentic, and prepared statistical data for scientific study and analysis. Access to federal data at scale could alleviate society's greatest problems by measuring needs and assessing ongoing government services. In proposing a national center, these boosters believed that adequate privacy safeguards already existed because government agencies aggregating data ensured confidentiality when information was disclosed by individuals and collected from them.

On the other side of the debate were alarmists—media pundits, legal scholars, and congressional representatives—who decried coordinated access efforts for fear of government overreach. They predicted the abuse of networked dossiers of every American citizen, using data collected from cradle to grave, that would bankrupt American citizens' privacy rights. As the publicity about the capabilities of automated recordkeeping and computers circulated, criticisms of the national data bank plan began to appear in newspapers and magazines warning of the invasions of privacy.[48]

To be sure, the three National Data Center proposals did not consider the abuse of individual privacy in their recommendations. Dunn would later admit that the gap surrounding personal privacy was a "gigantic oversight."[49] Despite the significance of planning for data access, boosters rarely raised limitations of *who* would be given access to the National Data Center or under what terms. For example, they didn't consider permissions for researchers or whether combined data could open new avenues for abuse or possibilities for protection once aggregated. Instead, more attention was spent worrying about realistic access services and format migration or whether there would be enough reference expertise and support for researchers and their queries. Once data was acquired and aggregated into a data bank, proponents had assumed that ensuring confidentiality would be a moot concern. But critics of the center assumed the opposite—that privacy would be placed at serious risk by aggregating personal information into population-level datasets, easily read by machines and accessible to all-knowing computers.

What the proposals did not say about privacy and the potential for abuse after aggregated data has been assembled for access reveals proponents' trust in government recordkeeping processes.[50] When scientists and research data managers were imagining what a social science data archive would be, they spent more time describing the processes of intervention between acquisition and possibilities for distribution than the original context of data collection and the impact it would have when used in research. Conversely, the public controversy over the proposals and criticism from congress promoted an increased awareness of the power in data extracted from files at the population level and the potential for abuse once digital data was aggregated for computational processing.[51]

REBUTTALS AND REALITIES OF DATA AGGREGATION

By 1967, social scientists began to publish rebuttals to the outlandish Orwellian claims in the media about the networked capabilities of data archives. They aimed to answer specific concerns raised in the congressional hearings on the possible invasion of privacy from a nationalized data center. Dunn, whose Bureau of Budget report was the focus of much of the hearings' scrutiny, published a point-by-point response addressing the alleged threat to personal privacy, and set out "correcting obvious misinterpretations" about the proposed center.[52] Many researchers claimed the hearings missed the point of accessing public data for scientific research, whereas a few said social scientists could easily do without data banks altogether.[53] Most scientists argued that in order to preserve confidentiality and protect the vulnerable, statistical information stored in data banks must be shared in a form that could not be "broken down" to identify individuals—which was something already occurring in most agency processes and enshrined into law.[54]

As social scientists began to push back on the sensational press coverage of the congressional hearings, nearly all made the important point that was lost in much of the testimony and press—that a statistical data center for research and administration would be much different than an intelligence data center intended for surveillance. But by then, the train had already left the station.

The American public was losing confidence in the government's record-keeping operations as more and more dragnet and surveillance programs were revealed during the 1960s.[55] In the data center hearings, legislators conflated the identifying information found in files that were aggregated into statistical data (e.g., age, race, gender, or income) from personal information stored in records and dossiers generated by institutions (e.g., high school graduation, a marriage certificate, or personnel records from military service). Questions from hearing members and testimony from privacy experts described the data stored in the National Data Center as personal information found in dossiers, not statistical data, which is anonymized once aggregated. In their questions to scientific experts, nearly all lawmakers conflated tabulated datasets of aggregated information with access to unique personal information found within individual files, despite the goals of these two types of "files" being quite different in their precision and scope.[56]

But the distinction between aggregated data used for social science and data collected for surveillance was never successfully argued by the experts who gave testimony. The rebuttals from most scientific researchers and data managers suggested that the maintenance of privacy requires consent and confidentiality at the time of data collection—that is, privacy is assured when the information is collected, not after being handled and processed into new storage formats or when it is accessed after aggregated in a dataset. The panel on privacy and behavioral research from the Executive Office of the President would also support this recommendation following the 1966 hearings.[57] To share information with government, subjects must meaningfully agree to consent to participate and be assured that their information would be treated as privileged and confidential. One scientist observed that the National Data Center debate "did not present new challenges, just concentrated the issue into one large package" concerning personal privacy and computerization in American life.[58] By 1969, most researchers and data bank managers publishing on the topic lamented that the distortions in the media would weaken the chances that a US National Data Center would ever come into being.

Looking back, we can see two versions of the problem of how to protect citizens' privacy using computing technologies that attempted to aggregate data to provide centralized access. Scientists trusted government assembled data to preserve confidentiality at its source or points of collection, whereas the public distrusted how data—once centralized and stored in machine-readable formats—could be recombined, programmed, and manipulated after it was prepared for access. In reality, the privacy and confidentiality that the lawmakers demanded in the hearings was already ensured with the siloed model built upon millions of stacked punch cards. But with the transition to magnetic tape and its increased mobility and programmable qualities, abuse would be more likely if aggregated records were easy to search, merge, and match. The only way to guard against such abuse would be through oversight, but ironically, such oversight couldn't be ensured without centralizing data sources from across the government into a data bank.

The congressional hearings featured expert testimony from many of the three reports' authors, including social scientists, government economists, bureaucrats, computer engineers, as well as legal scholars such as Arthur Miller and Vance Packard. Both Miller and Packard were early legal experts studying the legal frameworks and the privacy implications of new technologies such as computers and miniaturized surveillance equipment. Throughout the hearings, experts who had researched and developed many of the federal data center proposals struggled to distinguish the data center from a "total information system" that would support state surveillance systems.[59] Representative Gallagher was outspoken and argued that despite the potential for increased efficiency and cost savings, the National Data Center posed a number of threats to Americans' personal privacy, ranging from misuse to confidentiality violations.[60] Most of the data archive boosters and social scientists who followed the hearings argued that issues of confidentiality and personal privacy, although important, were "sensational and distorted."[61]

Researchers such as Sawyer and Schechter acknowledged that magnetic media such as digital data tape could be copied and manipulated more easily with computers. But social scientists argued that "privacy of recorded information has been protected in practice by the high cost of assembling records from scattered locations."[62] Whereas before data stored in paper records and punch cards had been prohibitively expensive to aggregate, they refuted that magnetic tape would make invading privacy easier because it was still labor intensive to duplicate and manipulate machine-readable data with computers.[63]

Alarmists such as Miller and Senator Long argued that the major threat to privacy results would come with the requirement that, to benefit from integrated records, each record must be individually identified and connected to a person. This point proved difficult for scientists and experts who testified in the hearings to communicate to critics because an individual record identifier was easily misunderstood as an individual's personal file, like an income tax return. For most economists and social scientists, there is no scientific interest in examining a single individual. Instead, they derive insights from aggregated data that rely on matching unique bits of information to a larger whole. As such, the requirement for matching to create datasets means that an individual record must be preserved to combine data in aggregate and to prove its attribution from the original source.

One bureaucrat who was very much in the spotlight in the Gallagher hearings in the summer of 1966 was Raymond T. Bowman, assistant director from the Bureau of Budget's Office of Statistical Standards. Bowman had previously argued that more problems could be alleviated if data collection was standardized and coordinated across agencies.[64] In testimony and in published articles, Bowman illustrated how statistical information couldn't be created without personal data about individuals. He believed that that the National Data Center would need internal access to such information, with the expectation that it wouldn't be published, and that the original agency had collected the personal information through legal means. Further, Bowman argued the abuses described in the hearings were already against existing laws for sworn government workers. Later, he would clarify that the Office of Statistical Standards wouldn't collect specific individuals' information and that some government records should be excluded. Bowman specified that personnel records, fingerprint files, medical records, and individual or business investigation files compiled by the FBI or other regulatory agencies would be excluded.[65]

Bowman was one of the few government experts who attempted to show how a federal center could both preserve privacy and provide comprehensive access using new machine-readable formats. He argued that there were numerous ways that data could be masked to protect identifying information during tabulation and analysis with computers. Like many other data bank managers with access to computers, Bowman believed that it would be easier to program and ensure the protection of identifying information with digital files because machine-readable files were less easily abused, as they couldn't be read by people easily.[66] Like other bureaucrats, Bowman argued that an all-inclusive information system wasn't technically feasible, nor was it the desire of any "responsible" government workers to break existing records laws to violate citizens' privacy.[67]

Whereas proponents of the National Data Center acknowledged the significance of individual privacy to American life, nearly all maintained that the criticisms levied against the proposals were largely misdirected.

Many decried the waste of the congressional hearings themselves, which created more public fear of potential overreach than understanding of existing realities of accessing data from data banks and archives. Surely, providing access to federal data would greatly outweigh potential costs of abuse from a repository that didn't yet exist. They believed alarmists conflated popular accounts of computer networks with misunderstandings of automated recordkeeping systems and their realistic deployment in government. Increased public criticism toward even *planning* to aggregate and coordinate personal information for data bank access wasn't just limited to governments but extended to industry as well.[68] Nearly all criticisms assumed that coordinated access to vast amounts of data would result in exploitation. In the words of one commentator, "a data center designed as an innocuous tool for statisticians would become a kind of electronic Frankenstein's monster."[69] By the end of 1967, scientists and bureaucrats had stopped lobbying for the National Data Center proposal altogether. The Nixon administration chose not to pursue the Bureau of Budget's recommendation, and research data managers began to focus on building professional organizations to develop standards for the patchwork network of data banks that were already established.

Toward the end of 1968, the Russell Sage Foundation commissioned board members of the National Academy of Sciences (NAS) to study data banks and issues of privacy in the US. The NAS research team was led by Alan F. Westin and Michael A. Baker.[70] For all the public spotlight and critical attention the National Data Center proposal had garnered during the hearings, there was actually very little empirically known about how data banks throughout the country were transitioning to computers, digital storage, and automated recordkeeping systems. Although "manual" data banks of paper records and punch cards had existed for decades, not much was known about how organizations were transitioning their recordkeeping practices from punch cards to new data storage formats such as magnetic tape and computing technologies. The goal of the NAS study was to generate profiles of organizations that were incorporating computers and automated recordkeeping into their operations. The case studies would gather evidence from educational institutions, corporations, nonprofits, and government agencies about how they were balancing personal privacy issues with organizational needs to access information. The studies specifically focused on current operations, aiming to observe transitions to automated recordkeeping systems, particularly in contexts where personal information was collected in files and then aggregated into datasets for secondary access.[71]

THE STRUGGLE OVER THE MEANING OF "FILES"

An individual's right to control personal information collected about them by an organization is always in tension with the purpose of why it was collected for the organization and how the organization aims to use it once it becomes aggregated. This tension between privacy and reasonable access still exists and can be seen today in struggles over the meaning of "files" in the National Data Center debates.

A file is a way to store and organize documents. It can be used to structure inscriptions and store them in information systems—organizations of files that organize documents allow users to link bits of information across discrete files. In the proto-digital world of punch cards and manual computing with tabulating machines, data had to be extracted and transcribed from structured records to punch cards. These records were human-readable documents like ones we still use today—hiring forms, school transcripts, resumes, and letters of recommendation. Most institutions that collected records about people bundled data in individual files. But as the transition to computerized data management diverged from manual data practices, the ontological distinction between human-readable information and machine-readable data collected and stored in American institutions came to the fore for government, science, and participation in public life.

What mattered for both sides in this fracas over files was what the original source of evidence captured and who guarded its access once it was transformed to other formats that machines and computer networks could leverage. Was it a personal file that could be rearranged and grow with identifying records to accumulate into an invasive "dossier"? Or would data be extracted from many different sources, becoming an aggregated dataset reflecting population-level trends? Both sides had visions of what would happen with computerized files once bits of personal information were accessed and stored in machine-readable formats. For both sides, the meaning of "files" centered on potential future uses. What mattered was not the adherence to the original source of context. Instead, the mutability and mobility of new forms of data stored in automated systems swiftly shifted the power of access to potentials of aggregation without oversight. This unforeseen deviation would spark a public debate over how the government could use citizens' information and how the public understood ownership over personal information and access rights.

Creating and sharing files about individuals is how modern institutions record and organize actions. But with paper records and analog file systems, documents were, to a certain degree, siloed and static. It took a great amount of effort to search, retrieve, and connect data across systems during the analog era of paper files and punch cards, all prepared by hand. With integrated circuits, faster memory cores, and new computing architectures, data could be more easily collected in digital formats. Combined with automated recordkeeping and standardized digital data formats, systems had the potential to be connected, even interoperable. For scientists and government researchers, computerized data management promised new ways to access aggregated data and build new bodies of information through centralization. More robust access and data management would surely allow economists, scientists, and government researchers to ask more questions and produce more knowledge.

But for congressional representatives and much of the American public, these new possibilities did not ensure unencumbered access and were more likely a harbinger of government overreach, increased surveillance, and the abuse of power. A National Data Center would close gaps between siloed data banks that were previously made of paper data and manual data practices, and in linking previously siloed data resources, the data center would ensure unknown uses and potential applications that had yet to be proven. For boosters, government centralization of machine-readable data for faster access services could only be good news, but for the critics, linking networks of computers to massive amounts of data collected from Americans was most surely a bad idea.

It is important to weigh the backlash to the centralized data proposals against broader social and political shifts occurring in the 1960s during the Vietnam War that began to shape the American public's trust in government providing access to accurate data. For nearly twenty years, the conflict made many Americans cynical and less trusting of the government, in part because of the data and information that was and wasn't shared with the public through news media. In 1967, the Johnson administration had launched the Progress Campaign, a public relations effort that circulated massive amounts of battlefield data to convince the media of more optimistic coverage about the progress being made by the US military.[72] By the time the Tet Offensive began in January 1968, Americans' support for the war had rapidly diminished, and "the credibility gap" coming from Washington was well known.[73] Americans became convinced that the numbers of causalities published in the media were being inflated to disguise a stalemate and that information about the war from the government was not the whole story. After Tet, it became clear that the data that the Johnson administration and military leaders were circulating to shore up public support for the war proved to be inaccurate at best and misleading at worst. Thus, many of the lessons from Vietnam such as Tet and Johnson's Progress Campaign showed Americans that much of the power in data collected by the government was in controlling how it would be used and for what ends.

The distinction between personal information and public data was conflated in discourse surrounding individuals' privacy rights, and ultimately who controls access guards the data use cases.[74] Most importantly, fears about the amount of information being collected from citizens that were uncovered in congressional hearings were "almost exclusively directed at the federal government, not corporations." Westin, Baker, and others explain that such problems of defining the purposes of data archives would then trouble data managers and researchers' proposals for centralizing federal data. Such definitional dilemmas opened the door for private institutions to continue collecting data about individuals with less scrutiny than the government's recordkeeping processes.

The public debate about the centralized data archive proposal rested upon an ontological distinction between personal information collected in files and how it could be used once aggregated in digital data formats for future access. During the transition to automated recordkeeping and networked computing utilities, personal information, privacy, and the greater scientific potential of accessing public data would become enclosed and entrenched in new markets. A conflict emerged in the transition from paper files to punch card systems and tabulating machines to data storage with digital tape and data processing machines. The conflict was about the ontological distinction between human-readable information and machine-readable data collected and stored in American institutions.

The debate over the National Data Center was also caught up in other social movements organized around the politics of free speech, bureaucracy, and government transparency over information collected at scale. By the mid-1960s, punch cards had become a symbol of both computerization of society and alienation from institutions increasingly using computers for recordkeeping and organizing knowledge.[75] For example, student protestors at UC Berkeley burned punch cards used for course registration just as draft cards were burned during anti-war protests.[76] Punch cards, in Fred Turner's words, allowed student activists to protest "the corporate world, the university, the military, and the punchcard [sic] universe of information" all at once.[77] The failure of proposals for a National Data Center and public outcry against the personal information the government collected about citizens illustrate early cultural expectations of what a data archive could be and who should have access to centralized data as well as who would appear in it.

Alarmists won out, ensuring privacy legislation and federal oversight for automated recordkeeping across federal agencies.[78] However, public fears about data integration and the collection of personal information would be a prophecy that would eventually come true, if not by the federal government then by organizations transitioning to automated recordkeeping and managing digital data in collections for access. Government overreach and abuse may have been curbed with laws that increased Americans' rights to privacy and limited the federal government's ability to match computer records, but the defeat of the National Data Center proposal also severely inhibited research data access and network computing resources of statistical data from federal agencies. Centralizing digital data for access, of course, couldn't be stopped. With the adoption of computers and digital data storage, the future boosters envisioned for the government instead came true for universities and corporations in enclosed private contexts that would create a market for personal data collection by third parties to thrive.

THE REALITIES OF AUTOMATED RECORDKEEPING

The public concerns and expert testimony in the 1964 congressional hearings about networked computing and the potential for data abuse were perhaps overwrought. Dunn and, later, the NAS report would prove that the Orwellian visions from critics that predicted "*1984* by 1970" were nearly impossible. Numerous studies of government agencies' data operations found that most did not have comprehensive information about data sources, storage locations, or formats.[79] Most of these congressional hearings and subcommittee reports—including reports from federal agencies and the National Academy of Sciences—argued that the invasion of privacy and potential abuse of data banks were not inherent in the transition to computerized systems and automated recordkeeping. Indeed, in their case studies of state and federal agencies, Westin and Baker found that there was no considerable change in the handling of personal information from manual data to machine-readable data systems.

Still, many legal critics' predictions and the public's concerns were sounder than social scientists and data managers at the time liked to believe. How personal information was collected in forms and then assembled, verified, and stored in files matters as much today as it did then. For decades, stakeholders inside and outside government lobbied for more information about the government's recordkeeping practices and data archives. Yet, well into the 1970s, there still was "no guide to the location of information being collected at the national level."[80] It is clear throughout their reports and rebuttals that experts and data specialists struggled to locate where the risks of accessing data in aggregate actually resided and where safeguards should appear in the data lifecycle to preserve privacy.

Many of the bureaucrats and public servants pushed back about the characterization of government overreach because current laws already prohibited the misuse of personnel records and individuals' information. They pointed out the significant differences between legal limitations of personal information collected for surveillance and intelligence efforts. Data collection from citizens was as necessary as it had always been to essential works of government for statistical analysis, economic forecasting, and enforcing laws that relied on aggregated personal information collected from states and federal agencies. But laws mandating accurate population statistics and data collection from citizens, like many others governing information and communication technologies, could not anticipate what digital data and networking technologies would soon bring.

Westin and Baker argued that the right to privacy and due process in the US was a publicly cherished value but that "the reality of the 1960s and civil liberties in record-keeping was an underdeveloped area of American law. Thus, computers moved into the world of organizational recordkeeping at a time when both American law and public opinion were just beginning to confront the problems of defining more clearly what rights of privacy and due process individuals should have in the various zones of manual recordkeeping that had come to exercise a major effect on the lives of individuals in the post-WWII era."[81] Debates about privacy through the 1970s would hinge on popular understandings of personal information stored in "files" as modern organizations transitioned to automated record-keeping and embraced computing resources.

Just a few years later, in his treatise *The Assault on Privacy*, which gave an account of the National Data Center's failure, Arthur Miller wrote that "the apparent victory is largely a Pyrrhic one."[82] Miller had been one of the staunchest critics of the proposal, publishing editorials and articles during the congressional hearings. According to Miller, the issue wasn't access to aggregated data after data archives had been amassed. Instead, he was concerned about how privacy was threatened in the processes of data collection and information handling that were being transformed with the transition to computer-driven recordkeeping—not only in government but industry as well: "Federal and state agencies were only part of the picture. Powerful private institutions that amassed, stored, and processed personal information formed the other part."[83]

Miller's point was significant then as it is now. During the hearings on the invasion of Americans' privacy, there was very little scrutiny of organizations outside of government that were manually collecting population-level data such as insurance companies, credit bureaus, polling firms, and advertising agencies. In failing to create a National Data Center, the government opened space for corporations to start profiting by aggregating data for their own. Preventing a nationalized data center enabled a new market for privatized data banks and data integration solutions to take hold and thrive.

Mid-century visions of accessing data for science, governance, and industry were tied to the potential vast caches of data collected by the US government. The government used data for providing services, for bureaucratic administration, and to enforce policies, but data banks across government agencies were siloed and manual data protected privacy by default. Not only would the transition from manual data to digital data influence approaches to data archiving, but computerized data management would also affect expectations for what computers could be used for and who could use them. Whatever side of the war on files and data archives one stood, everyone agreed that "the data cycle has new dimensions in the age of the computer."[84]

Just as research data managers and social scientists across the US were pushing for greater access to digital data stored on magnetic tape and punch cards at larger scales in data banks, computer scientists, hackers, and engineers in university research departments and engineering labs began to develop time-sharing utilities and operating systems for smaller computers to program and access data at greater speeds. Once manual data could be stored in digital formats, it would become more mutable—faster to process, easier to access, and more mobile. Data would still be made by hand, but once stored in digital formats for machines to process, it could become programmable. If the debate on centralization shows the stakes of data archives once data has been aggregated, then new dimensions of programmability and portability would be next.

3

MAKING DATA PROGRAMMABLE

The story of how data can be captured in a digital storage format and separated from originating contexts of creation is a process I refer to as "externalization." Today, data can be saved in flash drives and SD cards or easily bundled and transmitted wirelessly as text messages or Global Positioning System (GPS) coordinates. These processes assume externalization—data can be moved from where it begins, away from its origins. But this quality was not always so. To understand how digital data is archived today, we must examine the storage formats and technical processes that lead to externalization. This chapter shows that with processes of externalization, data has been distanced from origins of context through technologies of memory and externalized storage.

How does data become distanced from its original contexts? The answer has significance for how data is archived, but it also reveals a shift in the meaning of data beginning with mid-century approaches to data processing driven by computers. Early approaches to analog data management involved huge mainframe computers that used manual data processing and static media formats such as paper and magnetic tape for storage and egress. Data was never far from the machines used for processing it. But as mid-century computing technologies and data management developed, new grammars of action emerged with magnetic tape and time-sharing systems that allowed data to be moved and managed more easily. These external forces are embedded in the ways we save and access data today. Time-sharing, data processing peripherals such as tape decks, and interface terminals that incorporated magnetic tape for memory would change how data could be programmed and how fast it could be managed, stored, and shared across networks because of externalization. These began with magnetic tapes.

For most people, reels of magnetic tape conjure bygone experiences of listening to cassettes in the car or rewinding video tapes borrowed from the video rental store. There still are those who remember learning to interact with computers using tape drives with reels the size of pizza boxes. But few of us still own the cassette players, video cassette recorder (VCR) machines, or the computer tape decks to access the information contained in these tapes. Indeed, manufacturers have largely stopped supplying and selling them. Even so, magnetic tape is still used as the primary archival storage medium for digital data in cloud computing infrastructure and can be found in data centers such as Amazon Web Services (AWS) and Iron Mountain that offer tape-driven deep storage solutions.

Archiving data with magnetic tape isn't just about long-term preservation and data processing. The storage format also influences how data is created and made accessible through playback and representation. How archiving machines rely on formats such as tape or punch cards also reveals the stakes of accessing data once it has been stored and the means with which we can access it once data has been captured in or migrated to another format. This chapter introduces the memory and storage principles of minicomputer architecture to understand the power of distancing techniques—what I call "indexicality" and "closeness"—that resulted from the externalization of data to tape. I show how the uptake of this data storage technology when it was new would influence the spread of time-sharing systems and increasing interactivity with computers such that we expect to access data across contexts and at varied distances.

Computational processes that externalize data and make it programmable emerged with magnetic line tape that became popular with the rise of minicomputers in the 1960s and 1970s. Magnetic tape on reels allowed data to be more mobile and mutable than data from piles of punch cards. The ability to easily distribute tapes influenced perceptions of the circulation and development of computing resources, enabling people to share programs, datasets, and eventually what would be called software across both information and communication networks. Unlike mainframe computing with punch cards, early magnetic reels and tape decks used for access with minicomputers drew people closer to machines, allowing them to program data faster, more iteratively, and collaboratively.

Eventually, automated recordkeeping systems that used magnetic media formats would replace paper-based systems. Magnetic tape formats enabled individuals—not just government bureaucracies, corporations, and institutions of higher learning—to own, control, and easily share data that became increasingly personalized. These data management innovations that emerged and took hold with minicomputer architecture in the 1960s didn't completely replace mainframe computing storage approaches. Instead, the use of distributed minicomputers and centralized mainframes worked together to shape the two-tier access regime that we have come to know today. Platforms provide both cloud storage infrastructure and access to data archives for a premium as utilities, while users carry computers close, creating massive data caches but relying on corporate for access models to these networked services.[1]

Before minicomputers, mainframe computers had dominated postwar computing in corporate and government sectors for much of the twentieth century. Well into the 1970s, these massive machines took up a lot of space in many kinds of organizational environments, but few people interacted with them. Instead, most manual data processing happened by preparing punched cards and paper filing systems away from computer rooms. Imagine a large room the size of a high school gymnasium or the complete footprint of a building's basement. It is well lit, very clean, and filled with mechanical humming. You enter through locked doors. Now imagine a range of tall cabinets lining the edges of the room, not unlike the shelving stacks in a research library. But instead of books in these stacks, there are rows of glinting lights and flip switches on some cabinets, others have large pairs of black spinning reels, and there is a pungent aroma of ozone that smells like burning wire. At the center of the room, waist-high cabinets that look like top-loading washer machines hold packs of stacked disks the size of large, multi-tiered cakes. Nearby sit tables with huge mechanical printers that look like ancient typewriters with giant roles of paper falling to the floor. This dedicated computer room runs all day and all night, sorting sequences and processing batches of data, storing as much as 50 MB at a time. Controlling this computer means that only a few selected people are permitted to move to an "area of the room one walked through."[2] At the time, this mainframe computer was considered lightning fast, with vast processing capacity costing tens of thousands of dollars to lease and maintain.

Minicomputers were less powerful and had less memory than early mainframe computers locked away in clean machine rooms like the one imagined above. But they had unique advantages too. Minicomputers were less expensive and could support multiple processes with a variety of peripheral devices such as printers, and eventually these machines could be

Figures 3.1 and 3.2
Office configuration with people, PDP-11 minicomputer, and peripherals in an open office plan in 1970. Courtesy of the Computer History Museum.

Figures 3.1 and 3.2
(continued)

used by many users at the same time. Like a row of coolers that line the back walls of convenience stores, minicomputers were still the size of several cabinets arranged in stacks and weighed hundreds of pounds. They were described as "mini" because they were smaller than room-sized mainframes of the 1940s, and they could be outfitted for modern offices where more people could sit near them to interact and use them.

Unlike the high security and precise management of mainframe resources, smaller minicomputers could be managed more liberally: "they are usually run open shop—i.e. by the programmer himself, and the problem of turn around time are not so critical."[3] An early computer programmer argued that although the closed-shop model probably saved on processing time, open shops with minicomputers conserved resources by allowing multiple users to share the computer at the same time.[4] Minicomputers allowed more users to experiment with, hack, build, and maintain their own machines together. They were sold directly to consumers rather than leased to companies like mainframes from firms such as IBM and Remington Rand. The earliest of these open shops could be found in university engineering labs in the 1960s. Many hardware companies collaborated closely with research labs and university researchers funded by the government and would gift minicomputers for testing during the postwar era.[5] By enabling laboratory researchers and small organizations to own machines instead of leasing them, minicomputers in open contexts paved the way for more personalized computer interaction because users could develop customized applications.[6] By 1963, MIT's Long Range Computing Study Committee argued not only that minicomputers equipped with time-sharing systems and multiple consoles for access would increase scientific research capabilities of the university, but also allowing more people to share computing resources would benefit the institution's mission in teaching more students.[7]

Most importantly, before minicomputers and their open architecture allowed for customization, mainframes dominated approaches to processing data in centralized batches, primarily storing it in core memory and magnetic drums.[8] This chapter argues that magnetic tape reels used for external storage in these minicomputers would impact approaches to sharing data and customizing systems by separating data from machines, making digital data more programmable. Early minicomputers used in labs and universities would heavily influence the uptake and circulation of time-sharing systems. Time-sharing was not only popular with researchers and engineers located in universities and research labs. Once minicomputers with time-sharing came to campus, they became popular among the students who would soon graduate and work in other industry sectors. Further, open shops encouraged experimentation with programming and hacking machines, so that expectations of interactive computing systems to support data management would later appear in business, scientific research, government, and everyday organizational life by the early 1980s.

Magnetic tape reels mark the beginning of a grammar of action for storing digital data over time and space with machines because data could be externalized in a more portable storage format. The ways magnetic tape could be shared and the culture of access it shaped also point to the early origins of data becoming born networked. Eventually, long strands of durable magnetic tape wound tightly in compact reels made it possible for digital data to fit in your pocket. With tape reels, software and programs could easily be mailed, installed, duplicated, contributed to, and evaluated by loose groups of minicomputer users over many states and even internationally. Thus, loose networks of users would arise from the distribution practices of maintaining and updating software stored on tape.

Over time, portable digital formats became standardized so that editing, adding, or erasing allowed for a more real-time experience with machines and programming data among growing communities of users. New material components—such as double-sided mylar coating and redundant encoding features—also made tape more durable, allowing for new processes and more practical time-sharing. Once users could reliably share time and computing resources together in open shop networks, they would create coordinated practices of computing with many actors using the same machines. With personalization or communal computing, there came a need to distinguish and save actions between individuals that would lead to the categories of individual user accounts that we find commonplace today.[9]

As a data storage format, magnetic tapes were more reliable than the magnetic drums, early hard disk drives, and paper tape used in the 1950s. As tape drives for reading and writing data became faster, the ability to easily mount an operating system or run more than one program became possible. At the same time, the processing capacity of minicomputers expanded, and users' expectations of computers as shared resources and utilities increased too. Magnetic tapes could hold as much data as many hundreds or even thousands of punch cards, and they maintained their order better than looming stacks of card decks. Encoding data sequentially on a tape obviated the risk of a stack of cards falling over and losing their precise order, but data tapes were also cheaper and more resilient too. Tapes could contain more information and take up less space when stored on shelves than punch cards because they were lighter in weight. So, transmitting more data farther and faster became possible by mailing tapes through the postal service, allowing communities of developer enthusiasts to flourish.

A trade-off of this data externalization to tape is that reels' machine dependency for meaningful access relied on tape drives. Access mechanisms such as tape drives provide a means of writing or reading data representations onto a storage medium, and they use a combination of electronical and mechanical techniques to access data from storage formats such as tape. With magnetic tape storage, data must be indexical, encoded, and structured to represent meaning to both people and machines. But unlike paper cards, the human eye cannot see data stored to magnetic tape. Data's meaning would change, expanding to include easier access via portable, digital formats as well as accessible across time, but such access would become dependent on machines as they become more interactive. Curiously, interactive computing brings people closer to machines, creating more individual, personal experiences of creating data as a condition of externalizing and separating it from computers. Thus, although magnetic tapes themselves signify an early step in distancing machine-readable digital data, they also signal the rise of personal information through mobility and durability. Today's access regime of data archiving relies on distancing techniques that push creators away from data and pull them toward machines through features of human–computer interaction. In this chapter, I show how distancing—as an expression of power exercised through interactive features and computer architecture—began in a moment in computing history when data's material form enabled it to become programmable and mobile.

FROM CARDS TO REELS

Today, our computing machines and information systems feature graphical user interfaces (GUIs) that allow us to easily program data with clicks, manual gestures and touch, and even verbal commands. Although these interactions feel direct and in real time, many of the ways data is stored are enclosed, hidden, or distanced from us physically. Before interactive displays and digital formats such as hard disks, data sources were stored in punch cards and paper tape. Storage formats shape both the time it takes for computers to process data and the preparation of making data accessible to machines as well. During the 1950s and 1960s, most computing processes happened sequentially. Programs were prepared in advance, typically through punch cards organized in stacks or batches. Whether the task was tallying votes or performing calculations for payroll accounting, the machine would run each program by interacting with data sources stored in punch cards or, eventually, paper tape. This dominant approach to data processing was called "batch mode."

Batch mode in this era required relatively low interactivity between the user and the machine. Computers did not have screens or immediate feedback, and they were locked away in rooms with limited access. Instead, users had to hand over their card stacks to an engineer in a heavily guarded machine room and wait for the mainframe to complete the program before moving on to the next processing job. Tasks could take hours or days to complete. Such low interactivity from batch mode processing meant that the users had little real-time control over what would happen once their programs and data stored in punch cards had been inserted. Typically, these machines were shared by many different offices in large corporations or research units in institutions. So, computing resources were scheduled to the minute. Because there was little room for error when batch processing with mainframes, time away from machines focused on preparing batches of cards.

Batch processing also required individual users to dedicate time away from the computer for programming and assembling paper data—manually punching data onto formatted cards with keypunch machines. Batch mode also assumed limited access to a processing window of allotted time. So, a fair amount of attention was put into working with preparing stacks of cards in advance. In punch card computing environments, the punch cards *were* data storage, whereas memory—the processing power of the computer—was found in drum drives, tubes, and, eventually, hard disks. Punch cards were simultaneously data and storage media because they could also carry programs or directions for how to process the subsequent data. As such, punch cards, like tape, are both input and output media. Reloading a card deck for a program still required physical effort that prioritized the material order of the data stored in the cards as a stack and structured collection. An error would mean re-punching a card deck and reordering large stacks of cards.

Formats that allowed for the externalization of data, such as magnetic tape, brought people closer to machines for programming and processing it. Like the flash storage we use today, magnetic tapes were only readable by machines. Unlike earlier analog storage—cards or paper tapes with punched holes—the binary inscriptions made with magnetic pulses are not legible or perceivable to human eyes. As such, the externalization and programmability of data from machines is conditioned with distancing from using tape. In this chapter, I illustrate distancing techniques that can be found in magnetic line tape, dual tape readers (or tape decks), and minicomputing architecture found in machines from the Digital Equipment Corporation (DEC).[10] Founded in 1957 by Ken Olsen and Harlen Anderson, DEC was the leading minicomputer manufacturer in the computer industry until the 1970s. Throughout the 1960s, DEC released a series of minicomputers called "programmed data processors" (PDPs), which were more affordable than mainframe computers. Instead of calling their machines "computers," Olsen and Anderson chose to describe them as "equipment," and they sold components such as circuit boards and programmed modules directly to consumers. Olsen and Anderson noticed that interactive, programmable machines were more popular than large, general-purpose mainframe machines that didn't allow users to input data or receive real-time feedback.

In 1960, DEC began selling the first programmable data processor, the PDP-1. The PDP series would become one of the most popular commercial minicomputers over the following decade because it gave users the ability to write programs, develop their own operating systems, and modify machines with peripherals and software.[11] DEC's more open approach to modification and customization differed from the predominant mainframe business model. For most of the twentieth century, mainframe computing outfits such as IBM and RCA would lease machines and deploy engineers to maintain and fix them. Leasing agreements prohibited most mainframe clients from modifying or customizing the loaned equipment. Conversely, DEC sold their computers directly to consumers, offered a suite of different building-block components called "modules," and gave away manuals for free.[12] The modularity and suite of peripherals available for minicomputers allowed for unprecedented customization of programs and experimentation, and, eventually, the ability to link together machines as nodes across networks, increasing their interactivity by bringing more people closer to machines.

University departments, scientific research labs, and government offices came to favor DEC's PDP-6 and the PDP-10 machines for their customization and range of peripherals such as printers and teletype terminals. Unlike general-purpose computing jobs such as payroll processing or census tabulating, minicomputers could be programmed to run experiments, play games, and compose music. Later, memory from PDP machines could be hacked so that four to six users could share the machine at once.[13] Many computing historians and media scholars have explored the impact of minicomputer architecture and data processing from this moment through perspectives of specific machines at institutions, featuring the early hacking subcultures of sharing and swapping code.[14] But instead of focusing on specific machines housed in particular research laboratories, here I want to draw attention to the impact that magnetic tape reels had on the externalization and capture of data in a format that made digital data more programmable than storage common to mainframe computers.[15] The mobility of tape, as we will see, influenced the practical time-sharing and access to data across networks.

Magnetic tape and minicomputers such as those from the PDP series illustrate two types of distancing techniques that rely on the indexicality and closeness of interacting with computers. These techniques act as a tether to archiving machines; they pull and push users to and from data as they access it. Through "indexicality," digital data is externalized to tape but is not accessible without a tape deck and machine to process it.[16] Where before the syndetic structures of data in analog punched cards could be seen plainly by the human eye, magnetic tape's indexicality relies on a tape deck and computers to access, read, and encode or write new data to the storage media. A peripheral tape drive connected to a computer is needed to render data from tape legible to a human reader. So, this dependency between the storage media and the tape drive for access creates indexicality that is both material and symbolic for users who wish to interact with data.

"Closeness" is a second distancing technique that brings people closer to the computer as it processes and manages data for programming and storage. Because of machine dependencies, indexical data increases loops of interactivity between people and machines, bringing them closer to program and process data as they interact together. By shaping expectations of controlling and caring for storage media that hold data while making it mobile, mutable, and programmable, this second process closes distance by bringing users closer to machines for interaction. Users are drawn to the machine through routines such as mounting and feeding tape or programming data at the machine. Such closeness to minicomputers reduced the distance that divided users during mainframe computing with punch cards and batch processing while increasing interactivity with new data processing and time-sharing systems. Both distancing techniques influenced cultures of access and created new possibilities for creating memory practices, as we see through the innovation of magnetic tape and dual tape drives as access mechanisms.

MEMORY, STORAGE, AND MAGNETIC TAPE

In early computing environments, storage media such as paper tape and magnetic line tape were optimal for processing data at scale. Paper punch cards and tape were the primary form of storage from the late 1920s and remained standard until disk cartridges became affordable in the 1970s. Both magnetic and paper data tape were proto-digital data storage, but they were not computer memory in a strict sense. Transistorized digital computers with expanded internal core memory shifted how data could be processed and stored by creating an important difference between storage and memory. This shift would change the meaning of data in approaches to processing as well as archiving it. Some principles of basic computer architecture are necessary for us to understand the shift. But first, why do computers need memory and storage, and what's the difference between the two? Why does data need to be stored outside of computers for it to be accessed and then archived?

In computing, there is a storage hierarchy of components close to the main processor (known as the "central processing unit" or CPU). The CPU needs storage technology to accomplish tasks by bringing in new information, processing it, and sharing outputs. Confusingly, the CPU technology that is used to store information for processing is called "memory," but it is volatile and does not have much capacity to store information. When the machine is powered down, the data used in volatile memory is lost. The memory technology that is slower, nonvolatile, and persistent is known as "storage."[17] Early computers used cathode-ray vacuum tubes, woven magnetic cores, rotating magnetic drums, and transistors for internal memory, while punch cards or paper tape were largely used for nonvolatile data storage.

The first computer to use magnetic tape for storage was the UNIVAC I. Introduced in 1951, the UNIVAC I was the first electric computer, and it used a thin metal strip to process and store information to augment the machine's limited internal memory.[18] The UNIVAC's magnetic tape (called "Uniservo") had eight tracks for storing programs and data types. It encoded information onto tape in apportionments called "tracks," and within tracks, smaller portions of information were encoded as bytes in blocks across sections of tape. Recording density for tapes was measured in bytes per inch (or BP), and one roll of magnetic tape could rival the storage capacity of thousands of punch cards. Shortly after the UNIVAC was released, IBM developed the 726 tape drive in 1952 and the 10.5-inch reel, which became an industry standard. IBM's seven-track tapes were stored on open-faced steel spools called "reels." Similar to audio recording technology at the time, reels were mounted to tape decks, and the tape was fed to across to an empty reel, which would collect it as the machine processed the length of tape.

Tape decks for computers have two parts: a tape transport device that reads and writes information from magnetic tape, and a controller interface between the tape transport that transfers data between the tape and computer once the magnetic tape is mounted to the transport via two reels. Magnetic line tape—like ancient scrolls or wound objects such as a roll of thread used with a sewing machine—is stored tightly wound up on one reel and is then accessed by drive motors that pull the tape, unwinding the tape from one reel to another. When the tape is unwound across two spools, it lightly sits above a magnetic head between the two reels. The magnetic head can read and write information to the tape feed with magnetized pulses.[19] Early open reel tapes included an extension of twenty feet at each end of reel to be fed and clamped, keeping the tape taut across the head to prevent the tape from running off the spool or skipping during read acceleration. Tape drives have precise motors that can wind the tape from one reel to another, passing the head that can both read and write information to the magnetic medium. The head suspends a portion of tape in air, while posts in the tape head can read the bits and bytes on the tape tracks as it moves across the head.

In 1964, DEC introduced a smaller three-quarter-inch tape format called "DECtape," which was designed to be used as the main storage medium for a computer operating system.[20] DECtape could support bidirectional reading because of the unique approach to data encoding in which fixed blocks could be individually read and written using an address.[21] Each block would have a timing track that provided "every character a specific address, letting the user rewrite a single recorded character or even one bit in it without affecting adjacent characters."[22] Block encoding data in this way allowed tapes to be randomly searched, like a slow disk drive instead of serial access and sequential encoding. The format also allowed data to be replaced more easily because fixed marks from the timing track could be used to locate a precise position on the tape for information to be rewritten.[23]

Magnetic tape is known as a serial access storage format, where data is arranged physically on a sequential or serial axis. Because data was encoded sequentially in most early tape technology, it had to be read from beginning to end on a reel (and rewound to repeat). One novel feature of DECtape systems was how timing data for information was recorded to fixed positions on the magnetic tape. This allowed for a tape to be read bidirectionally, and time markers could be used to locate data read from the tape.[24] After the emergence of bidirectional tape formats, data could be randomly addressed and read at any place on the tape. Like reading a manuscript written on a scroll, random access was possible with precise motors because the tape could be rapidly wound or unwound across two reels to the correct point on the length of tape.

With dual transport tape decks, machines could also be configured to save or swap out files on the second tape drive, so as not to slow down programs on the main drive.[25] Both tape transports would have separate relay logic so that their speed and motion could be controlled at the same time.[26] Eventually, the DECtape system allowed up to four tape transports or eight tape drives for the PDP-11.[27] Later, DEC's time-sharing system would leverage expanded internal memory for asynchronous data input and output, with a stored library of programs to increase processing time. When used with the PDP time-sharing system, DECtape transports would also give users more control in switching programs to schedule and modify the pacing of time-shared jobs.[28]

This dual tape drive technology that allowed magnetic tape to be both memory and storage would prove to be an incredible boon for operating systems development, ultimately ushering in practical time-sharing with minicomputers such as the PDP-11. This advance didn't occur solely because of the DEC's dual tape drive design or the PDP time-sharing system, but it does illustrate how data was becoming increasingly programmable by bringing people closer to minicomputers with storage formats like bidirectional DECtape. By bringing people closer to machines, both the tape format and the access mechanism of the dual tape drive would shape people's perceptions of time for programming as a practice. As we see with the rise of the Unix operating system and its subsequent variants, the ability to

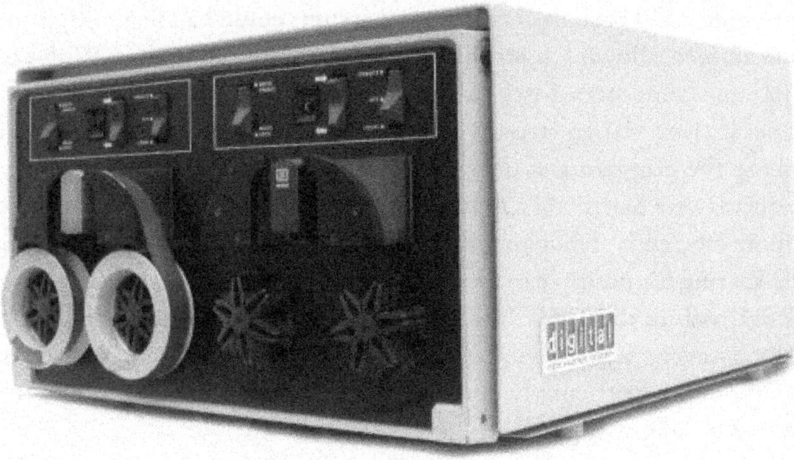

Figures 3.3 and 3.4
Digital Equipment Corporation TU56 DECtape magnetic tape drive, front and side views. Courtesy of the Computer History Museum.

quickly copy and distribute reels of tape that could be used across many different minicomputers for time-sharing would allow for data to be born networked among communities of users, from hackers and hobbyists to graduate students.

SHARING TAPES, SHARING TIME

Many organizations began to confront the limited approach to batch computing resources from mainframes during the 1960s, and business models began to develop around scheduling and leasing out processing time as shared computing utilities. Externalizing data also allowed for new approaches in distributing processing power instead of batch processing. As machines gained more capacity for memory, storage, and processing power, it became possible for multiple processes to be delegated, serving tens or even hundreds of users by allotting tasks and then swiftly changing to others and back. Increasingly, computers could handle multiple processes at once, allowing users to share "time" on the computer. With early mainframe computers, batch processing dedicated resources to one task at a time. If there was an error in a program, valuable resources were wasted running the program and deciphering the error, while sequential jobs in the queue were halted. As a result, time-sharing was first imagined as a way of more efficiently debugging and addressing stalled processes.[29] Although time-sharing for multiple users was first conceived in the 1950s and experimented with in the 1960s, it did not become common until the 1970s.[30] Multiprocessing with time-sharing systems allowed for more interactivity with machines, particularly between processing and handling the flow of data and programming it.[31]

With time-sharing, the computer is always running, but how it allots power to memory to process data with different programs can be coordinated differently, depending on storage. An early explanation of these approaches called these the "different philosophies of timesharing."[32] In one method, several programs are stored in different sections of memory. This works well for machines with large internal memory because different programs can be ready to be accessed by users whenever the computer is running. But for complex programs or for minicomputers with smaller memory, a second method that relies on auxiliary storage to store programs in nonvolatile formats is needed. When the machine switches from one program to another, the first in volatile memory must be swapped to external storage, while the second program is then transferred to internal memory.

Depending on the computer architecture and the method used to allot processing power to memory, early time-sharing systems ranged from specialized to general applications to the intended number of users being served.[33] Some time-sharing systems were dedicated or restricted to particular types of tasks, but multiple users could still submit tasks that were similar in purpose or structure.[34] By dividing a computer's processing power into smaller amounts of time spread across different input and output peripherals such as typewriters and printers, different tasks could be addressed in fractions of a second so that, eventually, sharing processing power among users became invisible. With dual tape transports and expanded internal memory for storing multiple programs, accessing data and stored programs could speed up. Time-sharing would collapse the time it took between loading a program, inputting data, processing it, and presenting the results to users. Without a distinguishable processing lag, such closeness resulted in a sense of "real-time" interactivity beginning to emerge, eventually becoming a design value for the personal devices that we use and expect from ubiquitous computing today.[35]

Although DEC and other minicomputer manufacturers had developed their own time-sharing software, customizable minicomputers such as the PDP allowed users to experiment with different approaches to data processing. Time-sharing with minicomputers came with the development of a new operating system called "Unix," invented at Bell Telephone Laboratories (BTL) and licensed to government and commercial users for a nominal fee by AT&T. The project was conceived and developed by Ken Thompson and Dennis Ritchie at Bell Labs in the late 1960s. Previously, Thompson had been a researcher on an earlier experimental time-sharing research consortium project called "Multics," and he had also developed a game called *Space Travel*.[36] Thompson took his previous experiences with Multics time-sharing research and the structure of *Space Travel* as inspirations for developing an early interactive programming environment for working with files. Unix was innovative because it brought time-sharing to minicomputers such as PDP machines, allowing users to manipulate files, create programs, and easily share their work. Early distributions of the Unix operating system were modified by university research professors, and their graduate students would customize it for particular use cases or implement improvements such as text editors, collaborative features, or game mechanics. Unlike the generic time-sharing systems created by corporations such as DEC and IBM for their customers, Unix was unique because it was nearly free.

Unix had a low-cost license for educational purposes, and it provided researchers in universities and labs new ways to develop new programs, customize their machines, and teach principles of computer science to novice users like undergraduates. Thompson had written Unix first in machine code for a PDP-7, and for a while, the software was developed and ran exclusively on DEC machines. Subsequent early versions developed at Bell Labs were for the PDP-11 and, later, its successor, the VAX-11 computer. But when Thompson and Ritchie recognized the possibilities of the system on a newly acquired PDP-11, they rewrote the Unix source code in the C programming language. By using the C language, any machine with a C compiler could install and leverage the time-sharing resources of Unix. The software soon spread because of its loyal users.

At Bell Labs, Thompson and Ritchie's time-sharing project was primarily intended to support word processing, but they had bigger plans for Unix. They envisioned a powerful operating system that would allow users to read, create, and interact dynamically with files together. In a personal history describing the motivations behind the development of Unix, Ritchie has said that he and Thompson "wanted to preserve [what] was not just a good environment in which to do programming, but a system around which a fellowship could form. We knew from experience that the essence of communal computing, as supplied by remote-access, time-shared machines, is not just to type programs into a terminal instead of a keypunch, but to encourage close communication."[37]

For the early developers and users of Unix, interaction was not simply controlling a computer, it was about communicating with other users through computers with this powerful operating system. As such, developing and contributing to the source code itself became part of building fellowship and community. Unix allowed people to interact and build community around systems and software because of the tape format in which the data and source code of the software was externalized.

The first licensed version of Unix was released by Bell Labs in 1969, but because AT&T was a regulated telecommunications monopoly, they could not profit from the software developed internally. Throughout most of the 1970s, versions of Unix were distributed to interested users on magnetic tapes sent through the postal service or hand-delivered from person to person. Early Unix source code was copied and shared on DECtape, DEC disk packs, and 9-track standard reels from IBM.[38] At first, there wasn't a formalized way of receiving a copy of the software. But usually, you'd receive a tape with the latest version in the mail directly from Thompson or Ritchie. An early Unix user reported that "BTL didn't really have a distribution policy in the early days. You merely got a disk with a note: 'Here's your rk05, Love, Dennis.' If UNIX crapped on your rk05 [tape drive], you'd write to Dennis for another."[39]

During a sabbatical from Bell Labs in 1975, Thompson worked as a visiting professor at the University of California Berkeley in the Department of Electrical Engineering and Computer Science, where he installed Version 6 of Unix. An earlier version distributed to educational and research organizations had arrived at the campus in 1973. Before Thompson arrived at Berkeley, he compiled a "diff tape" of Unix to share. The diff tape that he traveled with contained differences between the last version of Unix (V5) as well as a list of updates for Version 6. Many of these updates were bug fixes. So, the diff tape was also called the "50 bugs fix" tape. Early versions of Unix could stall machines or "crap" your tape drive because they had bugs. These errors combined with a lack of documentation and little formal support from Bell Labs left early Unix users frequently looking for support from other enthusiasts. Thompson has said that sharing the diff tape was significant to Unix users outside of Bell Labs because of limits from AT&T's consent decree: "I felt that it was important because I had discovered a whole set of very bad multi-programing (asynchronous) bugs that really should be fixed. The versions that were distributed were 'as is' with no provisions for updates. There was also some very strong language in the non-disclosure about users exchanging such information. The whole situation was unworkable and I felt that I had to do something."[40]

In Thompson's view, the importance of sharing the diff tape was that it would "get out" to other invested users, who could contribute the fixes and circulate more versions of Unix on tape, and indeed, that's what happened at Berkeley. Marshall Kirk McKusick, who shared an office with Bill Joy, recalls that "when Ken left to go back to Bell Labs, he left behind a Unix system. Bill [Joy] and Chuck Haley took over the care and feeding of that system. At the time, they did things like upgrading to Version 6 and incorporating the '50 bugs' tape from the Labs. Later they upgraded it to Version 7."[41] Both engineering graduate students, Haley and Joy began to contribute to the source code, adding text editing functionality among other useful updates that students could appreciate. This new Unix instance was called the "Berkley Software Distribution" (BSD). Just as Thompson and Ritchie had distributed Unix on tapes sent by post from New Jersey, Joy would send out free or, later, affordable tapes at a cost of $50–$200 of BSD from northern California.[42] For the next few years, researchers and graduate students at Berkeley would release four BSD versions of Unix.

As part of his distribution efforts of sharing versions of BSD, Joy would incorporate users' feedback to new versions of the system as he published and made updates to it. The first two releases of BSD enhanced Unix and fixed bugs. Over time, BSD became its own operating system, relying less on the original AT&T source code. As anthropologist Chris Kelty argues eloquently in his book on the cultural impact of free software, "the identity and boundaries of UNIX were thus intricately formed through its sharing and distribution. Sharing produced its own form of moral and technical order."[43] BSD Unix would later have an outsized impact on networked computing because it incorporated early Internet Protocols (IP) called "sockets" that allowed data to be read across early computer networks.[44] Having incorporated such networking protocols, BSD Unix became the most common operating system run on nodes of the ARPANET and, eventually, the early internet.[45]

The spread and uptake of software during the 1970s relied on the distribution of formats such as tapes and disks through postal networks of user groups at research institutions such as UC Berkeley. Unix matured and spread because its source code and updates could be easily distributed with tapes among a growing community of users. And as tapes such as the 50 bugs diff tape and versions of BSD were circulated, the system became more reliable and more widely used. Just as Unix was becoming more stable, different versions found on easily copied tapes began to take different meanings for stakeholders across an early network made up of Bell Labs engineers and lawyers, scientific researchers, academic computer scientists, their graduate students, and minicomputer hobbyists, among others using time-sharing systems.

Lou Katz, the founding president of the Usenix user group, described the fallout at Bell Labs once they discovered the diff tape had gotten out: "When the lawyers found out about it, they called every licensee and threatened them with dire consequences if they didn't destroy the [diff] tape, after trying to find out how they got the tape. I would guess that no one would actually tell them how they came by the tape (I didn't). It was the first of many attempts by the lawyers [at Bell] to justify their existence and kill Unix." For users outside of Bell Labs like Katz, circulation and access to the 50 bugs diff tape was a source of fellowship for the user base but an obstacle for lawyers attempting to enforce the AT&T license.[46] Or as Kelty explains, the circulation of an updated version was a problem for the lawyers because Unix needed to be stabilized for licensing purposes.[47]

Unix was developed on DEC minicomputers (mainly PDP-11s), but DEC had a frosty relationship with the growing and enthusiastic Unix community because it had developed its own time-sharing system.[48] As Unix began to replace more and more of DEC's own operating systems on PDP machines, the company reluctantly acknowledged users in the late 1970s by creating a formal Unix support group. As software like Unix spread and user groups began to form, networks of people began to share information and build publication networks to distribute software, documentation, and code. At first, these networks were built of engineers and specialists, academic researchers and their students, as well as computing hobbyists and fans. They were eager to circulate technical knowledge about computers and freely shared informational resources about software such as manuals, newsletters, and magazines.[49] In time, the vision of "communal computing" that Thompson and Ritchie envisioned for time-sharing through Unix was achieved through the care and contributions from other users at universities, research labs, and beyond. Loose communities of developer enthusiasts organized user groups, such as Usenix and DEC's support group, and built up new genres of close communication in publishing information resources, sharing source code documentation, and distributing data tapes.

The 50 bug fixes diff tape captures how data became programmable and more interactive not only through digital storage formats and dual tape drives but also through software distribution practices and time-sharing systems. The tape, as an early version of the Unix time-sharing system, would lead to community building efforts, resulting in newsletters and user organizations like Usenix. The circulation of the diff tape, copies of it (and other versions), was also a distributed storage strategy that would, in Katz's telling, work to circumvent Bell lawyers attempting to enforce software licensing and, in a short time, led to more and more student users in classrooms when it was used for educational purposes on campuses such as MIT and UC Berkeley. Time-sharing within the Unix operating system, sharing versions of the software such as the BSD, which contained networking protocols, and distributing source code through magnetic line tape show us the early stages of how data came to be mobile and mutable via programming and interaction.

By bringing people closer to machines for programming and access, minicomputers ushered in new ways of accessing core memory and leveraging external storage for time-sharing. These computing processes required the labor, routine, and physical knowledge of engaging and interacting with both data stored and the machines to render these formats legible for access and format migration. In focusing on the specific thresholds like the magnetic tape transport access mechanism where data is written, accessed, and stored to data tape, we can see how increased interactivity with computers came at the cost of new indexical constraints and structured dependences for representation and access downstream. It also shifted users' attention from control over managing data in cards to interacting with stored programs that would access and manage digital data formats. The transition from manual data processing with punch cards to magnetic tape that externalized data in a storage format that was easily handled and programmed using machines is a significant shift in our present conceptions of computing, interactivity, and processing data in community and networks. It is worth focusing on the shift in meaning of the word "data" itself.

Historically, "data" had assumed much physical labor in handling, punching, stacking, ordering, and storing data in paper cards or files. Manually preparing paper data for electromechanical tabulation processes meant that, in large part, people worked with paper information and data physically distanced from computers. As such, manual data was incredibly dependent on sites of handling and information transmission—from its collection and preparation to ordering and filing. Storage and memory architectures of mainframes also limited the time people used computers by partitioning windows of access. So, valuable computer time for batch processing also increased reliance on contexts of origination to define what was meant by data.

With formats such as tape, data could be quickly formatted, updated, and combined. Data had to be externalized in storage formats that machines could easily access and process. But by the 1970s, data switched to mean something more slippery and contextually independent because the material format and data encoding on the tape reel resulted in less manual handling and more computer processing. While relying on a machine to read, process, and represent digital information, data became programmable. Yet, the indexicality of data would tether its meaning irretrievably to machines for access from here on. Whether data was stored in a punched card deck, magnetic tape reels, floppy disks, or a spinning hard disk, the grammars of action related to storage and externalization started to involve indexical dependencies that would shape people's perceived closeness to machines and real-time interactivity in encountering data.

Dual tape systems such as DECtape and the mobility and programmability of magnetic tape led to the distribution of software and time-sharing systems like Unix BSD, driving public expectations of more interactive computing with shared resources. Although tape drives increased time-sharing capabilities and tape itself could be easily shared via post because of its size, the inherent indexicality of machine dependencies from tape drives were not without drawbacks for accessing data later downstream. The trade-off of more interactivity with computers is that interaction is dependent on machines. With externalized storage formats that are inherently digital, making sense of data relies on machines, as human eyes cannot read the magnetic inscriptions on tape reels without a tape drive connected to a computer.

Such machine dependencies on data access impact not only its representation in early circulation but also possibilities for long-term access if data is to be archived. Long-term archival preservation of data ensures reliable access to authentic, original forms of evidence, including hardware playback capacities for software-dependent media. For example, if we were to locate archival copies of a DECtape with Unix Version 6 and an IBM 9-track tape with BSD, would we be able to meaningfully investigate the updates, bugs, and features that are present in both? It would be unlikely for several reasons connected to machine dependencies and the indexicality of the data storage format. First, drive compatibility is a well-documented preservation access issue with tapes from minicomputers. Although magnetic tape held more data more affordably than any other data storage media until the 1980s, few versions of tape were interoperable across hardware until the late 1970s.

Reels of tape were cheap, but then, as they are now, tape drives required for access were expensive. Without a proper, functioning tape drive that matches a particular tape format, data cannot be accessed, rendered, or made useful by machines of different types. The same is true today as it was then—historic magnetic tapes are inaccessible without functioning tape drives and compatible machines, which have become rarer over time. Such dependencies are countered with representations made up of interoperable standards called "metadata." Metadata makes data meaningful and ideally accessible to both people and the machines. But like any structured system, metadata is a cumulative process that builds up over time by accounting for and mitigating access constraints. For example, a challenge of tape from these early years was that the metadata documenting the context and content of the tape, if it was present at all, was often separated from the data itself.

Despite the durability of the tape composition, most DECtapes used with PDP machines cannot be read and accessed now because there are so few PDPs left in operation with functional tape drives. Restored minicomputer exhibits and emulators exist, but during restorations, curators and collectors report that complete data tape preservation is not possible without readable documentation that describes the context and content of the data stored on the tapes. Although many archives store these tapes, there are limited resources for tape playback.[50] As such, the original metadata, separated or absent, makes most tapes inaccessible. Instead, to tell the history of software like Unix BSD or hardware peripherals such as dual tape drives, computing historians and archivists rely on oral histories, interviews, and firsthand accounts from users themselves. These interviews can then be supplemented with equipment manuals, gray literature, and documentation created by early user groups such as listservs and newsletters.

The externalization of data with digital formats not only impacted the meaning of data but also shifted what it means to access and then archive it with nonvolatile storage. Today, modern data centers continue to use tape as their primary data-archiving storage format. The medium's stability and durability has ensured its longevity. Because tape can be duplicated and safely stored on shelves, it allows for air gaps between interconnected, online systems.[51] It also supports long-term retention requirements from business communication laws and policy that govern how long corporate records should be retained. Tape density and durability still outpaces hard disks, which, combined with its affordability, makes it the primary medium for contemporary data archiving.

However, it is important to consider how data access with tape and time-sharing systems generated new dependencies for managing data collections, such as expectations for storage as a computing utility that results in time-sharing privatized storage infrastructures. After decades, design trade-offs from unique tape formats and tape drives that weren't interoperable across platforms created new indexical dependences. These dependences were pegged to the lengths of computer product lifetimes instead of favoring human legibility and long-term access.[52] Unlike the proprietary formats of the past such as DECtape, the tape technology used today in cloud storage benefits from open encoding standards and storage consortiums of vendors like the Active Archive Alliance or the Linear Tape-Open Consortium. Now, indexicality that increases our interactivity machines also leads to new systems of representation such as metadata. But for early uses of magnetic tape, the absence of metadata means that most historic tapes cannot be accessed via playback, whereas most new data archived today is stored on tape and automatically generated with metadata for access. On its face, this is a curious archival contradiction, but as I show in the second half of this book, it is inevitable in an age of automatic saving and platform-driven models to data access.

Data has always had theoretical enframings that alter or condition its reception as it takes form. Processing and handling data in the externalized form of the magnetic tape reels—as storage for early minicomputers and as open architecture—brought more people to computing. Once data stored on magnetic line tape became external and portable with reels, it made users' access to source code, as well as their ability to program machines, more direct. These grammars of action would later influence our perceptions of the mobility and transmission of personal information, as well our expectations about who can exchange, distribute, and own it today.

By 1970, MIT and DEC had developed file systems for data tape like DECtape. Files could be arranged in "hierarchical structure of up to forty levels."[53] These early tape structures for data files were used to represent data to both machines and the people programming them and, in time, would influence approaches to managing data and its representation. Thus, programmable data creates a new valence of meaning for data as it is externalized: the indexicality of machine dependencies for access results in hardware dependencies but with representations of data with structured systems such as file directories. File systems for data tape would have consequences for representing data and accessing it with shared computing resources such as time-sharing systems, but soon it would also shape how data would be managed by individuals once everyone had access to a personal computer. Data had not yet become personal before it could be managed and represented easily with software and storage formats.

The power of archives has always been in their organization and control. Whoever has access to archival collections gains enormous power. They acquire the power to speak with it and to make evidence, knowledge, and memory. Once digital data could be archived by computers, power would then be located in how data is structured, rendered, and represented in systems to both people and machines. When data leaves sequential formats and manual preparation like punch cards for digital formats like data tape, interactive systems are needed for both humans and machines to read, manage, and process it.

It has been said that in the 1970s, computers came to the people, but more accurately, new architectures and open shops brought more people to computers than ever before.[54] Processes such as time-sharing, magnetic line tape for storage, and software distribution networks allowed select individuals to interact more with computers for communal computing experiences.[55] In time, with higher interactivity, minicomputer users especially began to experience—and shape—grammars of action that relied on the increasing externalization and programmability of data.

Data stored on reels of tape then allowed growing user communities with access to cheaper and faster minicomputers to develop relationships through the development of source code and the distribution of it. In time, these new relationships to externalized data and interactive time-sharing systems would present a challenge to public understanding of how to benefit from data collected in the growing data banks of the 1970s. Communities of software users would shift too, moving away from tinkerers, hackers, and hobbyists of communal computing to mainstream vertical markets of software products alongside the growing personal computer market for hardware directed toward individuals. With this shift, the moral and technical order of data management would become more personalized, and with personal computers and devices, individuals could create more data faster. But they would soon become more responsible for managing it too.

Today, the consequences of making memory mobile can easily be seen in contemporary media formats such as flash drives, CDs, and cloud storage. But less obvious is how the portability of magnetic storage media began to shape the significance of personal information and software development in the 1960s by shifting perceptions of how data could be owned, accessed, and networked. By externalizing data to a portable digital format, the meaning of data itself would change from something handled and sorted by machines to something dependent on computers to be processed and read altogether. Data processing with digital tape would eventually create a space for the emergence of a new category of "personal information" that was mobile, mutable, and durable because it externalized data in a digital form that could be easily mounted and transported with software. As individual users interacted more and more with machines to process and program data, accessing information with magnetic storage media through computers would become more of a personal experience as well.

4

DATA OUT OF POCKET: FROM FILES TO APPS

Computing historian James Cortada identifies the 1980s as the beginning of the "PC era." The first personal computer users had a need for specialized software and central data processing. Sensing opportunity, the computing industry began to develop software applications that could be bought off the shelves and easily run on a home machine. Modems also started to allow legions of new users to increase their networking capabilities. By the early 1990s, PC users demanded access to large electronic filing resources, like computers of the 1960s.[1]

Today, most of our personal information is managed with automated processes, and the management of our personal data is typically outsourced to third parties such as data brokers and information technology firms. Practices of naming, sorting, organizing, and keeping files in directories have now become subsumed and automated by operating systems, apps, and platform services like Spotify, iPhoto, and Google Drive. At this point, data created by users is tacitly extracted and pushed to the cloud where most personal information is automatically managed by software. As a result, today's computer users have less experience with the personal information management practices of early digital files and directories, like analog files and folder organization of the paper systems from decades prior. The PC era brought together new consumers with interactive experiences driven by data and networks. In the process of being formatted and gathered for third-party access, data acquires new capacities for representation with automated metadata and network transmission. This evolution expanded what data archiving practically means and what meaningful "access" has become for different stakeholders.

Earlier chapters addressed how digital formats and computerized data management techniques impacted people's awareness of personal information collection on behalf of institutions. The next half of this book turns to the transmission of data and the various means of accessing it once it is collected. This changed who is responsible for managing data and how expectations of access change once personal data management is supported by more stakeholders and networked technologies.

This chapter shifts our attention to the personal information management tools and synchronization software developed by tech companies for portable devices during the PC era. Specifically, I detail how data management began to shape individuals' personal information practices and increase their willingness to exchange data and control for wireless software services and computing utilities. In the late 1960s when Congress debated the stakes of sharing collected data across agencies, they found the proposals for massive data aggregation shocking and incorporated safeguards to delimit machine-to-machine reporting with the Computer Matching Act.[2] Despite Congress's prescience, by the 1980s, collecting and aggregating personal information with computer-supported recordkeeping systems were commonplace business practices in the United States. Matching data from Americans' records and information became a way of classifying and organizing actionable computer consumer surveillance information for purchasing predictions, trend forecasting, financial services, and targeted advertising campaigns.[3]

The first portable computers in the late 1980s included "luggables"—terminals that looked like bulky suitcases—and tablets with stylus inputs called "pen computers." Despite the novelty of a stylus that allowed for faster data entry than a full-size keyboard, pen computing with tablets remained unpopular with consumers. Tablets were tedious to enter data and had crowded interfaces that limited their utility. Instead of being targeted to individuals working at home, most of these early tablets were marketed to vertical industries where computers could be used to support work "in the field." Ideal for supporting package delivery, managing supply operations, and entering government and enterprise data, tablets with pens were adopted wherever workers needed real-time data input and access to structured resources and databases such as inventory lists.

Unlike rugged field tablets, early portable computers were made for the home.[4] Individual consumers were attracted to promises of interactivity and specialized software applications. The first personal computers like the Atari Portfolio were about the size of VHS tapes and had built-in software such as text editors, address books, and spreadsheets. But the power and memory limits of early portable PCs constrained their portability. They required a wall outlet for power, two-tier memory storage, and expansion cards to run more complex operations like a chess game or to launch a file manager. Unlike most workstation computers in offices that allowed users to communicate with each other with email or intranet systems, first-generation portable PCs were difficult to hook up to networks. Transmitting data took a lot of planning and coordination. In Kevin Driscoll's words, "moving data from one machine to another was one of the fundamental challenges of early personal computing."[5] Most portable PCs could not communicate with other devices off the shelf. Networking was only possible with an expansion card slot and a modem that would turn portfolio machines into small computer terminals.[6] Despite their limitations, small digital devices continued to multiply, bolstered by consumer appetite for the burgeoning personal electronics market.

By the late 1980s, most digital handheld devices such as graphing calculators, language translators, and digital diaries supported only one or two dedicated tasks. These affordable devices had a form factor that consumers wanted; they were small, battery powered, and easy to carry. However, the operating systems and software applications designed for portable PCs and tablets offered too much computing power for early, novice users. Most people still used paper to organize and store critical information at home and at work. Even with a stylus, data entry on early digital devices couldn't compete with the ease of handwriting with pen and paper. Most workplace professionals relied on day planner paper organizers for managing work tasks and personal information until the 1990s. Despite their cumbersome form and lack of adoption, early digital organizers presented a vision for the future of mainstream portable computing. Smaller devices supported with software applications could allow a user to quickly move between dedicated tasks, creating and storing data.

Enter the personal digital assistant. The phrase was coined by CEO of Apple John Sculley at an event in 1992 announcing the Apple Newton, and the PDA would go on to change how the computer industry brought new users over to using personal computers and digital information. When Sculley announced that the future of personal computing would be with PDAs, there weren't many specifics. Many early accounts of what is now characterized as "mobile computing" centered on Apple's Newton Messenger PDA. Hype that circulated about its sleek design and AI-powered handwriting recognition software promised it would be the portable device

Figures 4.1 and 4.2
“Luggable” Gavalian SC Mobile Computer (top) and Atari Portfolio (bottom). Courtesy of the Computer History Museum.

to solidify Apple's legacy in the new market of devices that could manage your personal information "on the go." But despite fanfare and promises from Apple's CEO, the Newton was a dud. Less powerful, more utilitarian devices like the Zoomer beat out Newton, but not in the ways people predicted. The most popular PDAs couldn't support users communicating with others across networks easily. Instead, the Zoomer and, later, the PalmPilot allowed people to carry data in their pockets. These PDAs enabled them to easily synchronize data across machines using early software applications, much like "apps" found on mobile operating systems on smartphones today.

Early PDAs even looked much like smartphones with basic apps, minus the robust wireless networks and communication features. Most people today looking at a PDA would probably describe it like a clunky, ancient smartphone without Wi-Fi or mobile networking. Early predictions of PDAs focused on personal information management tools and features that met other personal information needs that were increasingly becoming digital, like communication tools and reference information. Marketing materials from the first PDAs emphasized their ability to help people send and receive data in several different formats, ranging from organizing their daily schedule to collecting and updating contact information and accessing email. They allowed users to create notes, sometimes with digital handwriting, and keep them alongside other files, allowing users to read documents such as emails, faxes, and eBooks. But ultimately, PDAs were strictly designed for a one-way personal experience.

Although PDAs focused on personalization, there were not intended to stand in or replace stationary personal computers or portable but heavy "luggables." Instead, the computer industry anticipated they would open a lucrative new market for helping consumers access and manage their digital data with smaller portable devices. For example, interviews with CEOs and product developers at Apple, Sharp, and Casio in the early 1990s emphasized that PDAs would help people consume information and not just create it.[7] How exactly information flowed to and from these early devices—without a network connection—is pivotal to understanding how they impacted the realities of archiving personal data today.

Early handheld PDAs that emerged in the 1990s failed to gain mainstream adoption because they had limited wireless networking. Even so, I argue that the control they gave to users to manage their personal information

Figures 4.3 and 4.4
Newton MessagePad had basic apps and handwriting recognition. Courtesy of the Computer History Museum.

with software apps presaged the storage experiences and data extraction realities we face today. Before wireless connectivity was an assumed feature of mobile computing, devices such as the PDA began to influence record-keeping practices that were increasingly automated and networked. How users were expected to move, access, and create data with early cell phones and mobile computing devices laid the groundwork for distancing data at scale today. Like the experience of using minicomputers such as the PDP, synchronizing data across multiple devices, even if the stewardship was tedious and time-consuming, elicited feelings of closeness. Later, the practices of synchronizing one's data would be supplanted by hidden seamless updating over wireless networks, ostensibly to save people time.

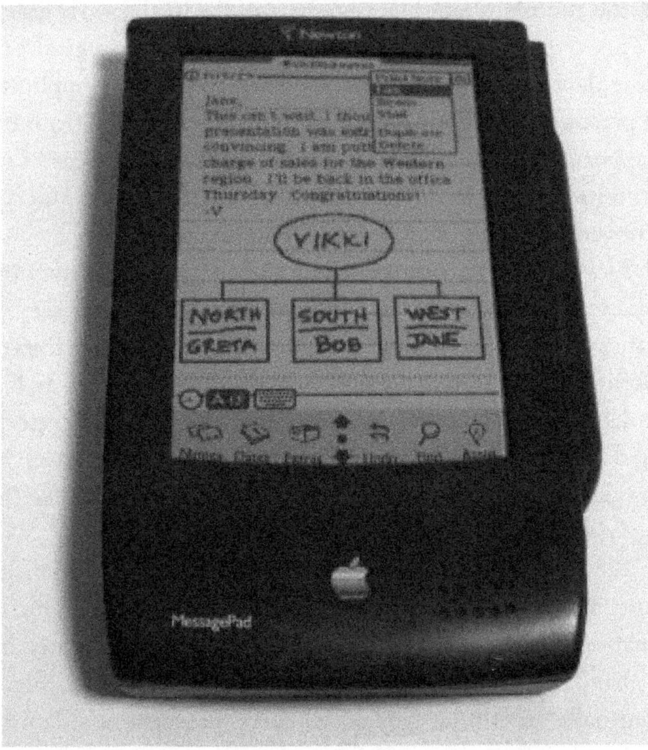

Figures 4.3 and 4.4
(continued)

Today, information is represented on our personal digital devices not through files but instead with the frame of apps. Mobile apps allow us to access, stream, and transmit data without considering the format of information. For most wireless users, apps are the key to accessing networks of information resources, communication technologies, and modern personal telematics. However, they also connect users to a wider ecology of stakeholders such as firms, software developers, and brokers that extract data from users through the same networks that support wireless transmission of information.

This chapter shows how the abstraction of the file and directories carried constraints from an earlier information order to the next.[8] From paper structures to data architectures to apps, I show that data traveled across earlier generations of portable computing devices and software that largely resembled the mobile operating systems we use in the early twenty-first century. This transition from files to apps began with digital organizers aiming to replace paper day planners and early software applications for managing personal information. However, as apps emerged from PDAs and began to appear on early cell phones, files began to disappear as representations of containment. As devices shrank to fit in pockets, data could also easily be moved across digital networks—no files required.

Indeed, files still exist as representations of structure and containment in our information age, but only up to a point. Apps changed users' relationships to data transmission and understanding of them as containers.[9] Data's transmissibility and mutability only increase as files disappear into apps. So, when data can easily move in and out of pockets, the conception of containing data in files within directories breaks down. Likewise, the rise of apps alongside the breakdown of files as containers has stakes for the representation of data and who controls it.

By the 1980s, the file, as a representation of structured and formatted data, became a way of understanding and organizing both digital and analog information. Once personal workstation computers appeared on workers' desks, handling and managing information using digital files became habits, eventually becoming part of workers' professional identities. The grammar of action from files and filing practices came to influence digital recordkeeping not just in scientific research and bureaucratic governance but also in business, commerce, and home entertainment. As these information practices became digitized in many sectors of society, data management too became personal. Data, as it came to be represented as files on personal devices such as PCs and PDAs, had to be managed individually.

Once these personal devices could connect to wireless networks to support data transmission, managing it would shift to automated processes, transforming the means of representing data as well as who can archive and access it. But before data could be transmitted across networks, machines that connected to them had to fit into people's pockets. As we will see, files as representations of containment would break down with new network processes of data synchronization and mobile transmission. Managing data with apps on mobile devices would impact the political economy of personal information by opening data to be easily copied, moved, and changed. Data would become managed not just by personal computer users or those who shared computing resources at work. It became controlled by third parties well beyond the original contexts of data origination and transmission.

As discussed in chapter 3, automated recordkeeping established in mid-century state and scientific computing dramatically changed by the 1970s because more and more American workplaces were becoming computerized.[10] Bureaucratic administrators still used paper records and manual data analysis for governance and administration, but social scientists, economists, and computer scientists began to collect heterogenous data of human activity at population level scales. These computers and processing approaches allowed for shared utilities like time-sharing for working with larger datasets while collecting information at smaller scales outside of traditional institutional contexts.

Alongside substantial technical changes in the externalization of storage and faster processing, recordkeeping systems spread data management practices to new areas of society. But digital technologies did not replace paper systems wholesale. The mid-century, paper-based analog systems made up of file structures and filing practices were still used alongside computers. Such paper systems would continue to influence how people understood information, even when using faster, smaller machines to collect and keep data. However, new magnetic formats, such as tape and floppy disks, allowed data to become programmable and rapidly processed. Each new format inherited grammars of action from previous legacy data management systems. With each new iteration, the meaning of data evolved. As such, the roles and responsibilities of the people designated to manage it changed too.

After minicomputers and time-sharing systems became popular, computing machines began to shrink, as they served fewer people and did more software-supported tasks. Rather than computers having a dedicated room you visited to reserve processing time, they turned into large office equipment. These smaller, more powerful computers began to be regarded as organizational information technology, supporting individuals in organizations at their desks and workstations with designated tasks.

We have seen that personal information was collected at scale by institutions for government bureaucrats, social scientists, and university administrators. But when did individuals begin to handle their own data? When did personal information become managed *personally* with computers? The answer to these questions lies in how smaller machines were introduced at work and dedicated to software that supported individual users. In time, these tools would become a new category of "personal information management"—a suite of software applications designed for personal and not organizational data management. These "apps" would rely on file representation of personal data but would also weaken it. Ultimately, apps built on wireless networks would restrict access to data archives, still experienced today through data extracted in the service of transmission.

In part, the wireless infrastructure of data transmission and harnessing networks is a technical story about how the history of data management led to today's cloud computing architectures. However, I argue that it is also a result of personal information management practices influenced by mobile computing devices and communicating with early PDAs from the 1980s and 1990s. A new paradigm of creating more easily shared content on networks came at the cost of individuals managing their own personal data. Off-loading personal data to third parties for management—initially seen as a boon to users—would shortly curtail users' access to their own data.

GETTING PERSONAL: PCS, DATA, AND PIM TOOLS

Practices of measurement and prediction based on recordkeeping systems have been around for centuries. With the rise of minicomputers and automated recordkeeping from the 1950s and 1960s, we learn how information in files and programmable data was collected at scale. Technologies such as time-sharing and tape made data mutable, easily copied, and shareable. But the representation of digital information as files also allowed for the state, private companies, and individuals to structure, store, access, and organize data from individuals for different purposes. But lax data protection policies from corporations in the US and the personal computing market enabled the massive collection of personal data that populates the privatized data archives assembled in platforms today. But as the idea of files as containers expanded from cards to tape, our understanding of representing data would slip in surprising ways. The file as a container became fuzzy. With machine dependencies for access and storage, ownership would shift from manual stewardship and custody to transmission and access via software. Once personal information management expanded to include data from mobile devices, containment was no less apparent to those controlling it, but post-file PIM tools involved more stakeholders managing access. I contend that before digital data could be archived by third parties such as platforms, the containment metaphor of the file had to both hold and fall apart with PCs and PDAs.

The file may appear to be a durable abstract representation of information. We still attach them to emails, upload them to cloud storage, and share them in group text messages. But files are tricky abstractions that are subject to both social and technical processes of the present and to grammars of action that reach back in time. If we return to early mobile computing devices like the PDA and cell phone or even to the communal computing era of minicomputers, the file abstraction becomes unstable.

In mental models of data management, files during the late 1990s and early 2000s were still used as containers for digital information. But like an accordion that can be pulled out and extended, how data is contained and physically managed has changed with transmission processes such as synchronization and wireless networking. Synchronization and wirelessness rely on stacks of infrastructure and coordinated arrangements between stakeholders such as hardware manufacturers, software developers, cell phone towers, and internet service providers creating automated transmission data that can no longer meaningfully be represented with the file to users.[11] The app now appears to stand in for the catch-all container for data today.

Files on early PC devices and shifting ways data was understood would enable apps to connect us to the ecology of stakeholders we only "interact" with via software, stacks of code, and wireless infrastructure. But the presumption that apps simply replaced files as containment abstractions of structured digital data is dubious. Instead, apps become access points. With early PIM apps on PDAs and cell phones, we can see how the file would become taken apart, atomized into layers of data and metadata that could easily flow into data archives, distanced and inaccessible to the individual data subjects who contribute to them. Still, the representation of information drove this shift because it assumed a prior abstraction of containment in files but leveraged the digital mutability and speed of wirelessness transmission in packets.

In the past, data were more valuable once structured and collected into files that could be easily organized, classified, and compiled for measurement. For much of the history of early digital files, their systems aimed to be both human and machine readable even if driven by automated processes. The aggregation of data that scientists and government bureaucrats lobbied for in the 1960s would become possible with personal computers and wireless communication networks leveraging standardized formats and transmission protocols. In time, files would become valuable when they could be linked and transformed into aggregated data. Once transmission networks connected, a new layer of representation would challenge access because the future promises of accumulation and archiving made data ever more valuable.

Small software applications that allowed users to interact with digital files on handheld personal organizers and PDAs enabled the rise of personal data management. These new, smaller devices and their early "proto apps" were PIM tools that relied on the understanding of files and folder directories common to personal computers as well as paper organizational schemes. But with PDAs that used apps, the file became a leaky abstraction of structured information. In other words, many early computing operating systems straddled analog and digital semantics to communicate and signify practices of organizing, keeping, finding, retrieving, and storing documents. Magnetic media represented data using concepts drawn from paper-based filing systems.

With apps, "the file" has faded across the ecology of wireless, mobile devices that represent contemporary ideas of data transmission. Instead of gathering information by structuring and storing it to files, early mobile devices shifted practices from data entry to synchronizing across machines. A later shift from containers to syncing data across networks broadened the constellation of stakeholders who created, organized, and accessed information. New opportunities emerged for asserting control over collections of personal information and data managed by corporations that own and govern stacks of infrastructure that allow data to be created, transmitted, and collected.

This destabilization of "the file" as an abstraction in favor of "the app" reveals how asymmetric access between third parties, government, and users became established. The weakening of the file, which began with information management for individuals with personal devices, would eventually be fully supplanted by automated systems that underwrite most apps today. But the automated personal information management that apps provide is best understood as a sleight of hand that obscures the significance of data management automated with apps and network-driven devices. To understand the significance of apps to personal information management, we need to further explore the history of early portable devices that would become mobile computers.

Mobile handheld devices have been used since the 1950s to create and process structured data. However, most early devices, such as calculators, didn't persistently store information. The new personal computer market that developed in the 1970s and 1980s triggered demand for personal information management software both at work and home. PIM tools began to appear in both personal computers and small consumer devices called "digital organizers," which supported scheduling appointments, writing email, and keeping address books. As computer components began to shrink and drop in cost, engineers and designers started planning for digital devices that could easily be held in your hand. They would have enough storage and power to handle all your personal information *and* fit in a front shirt pocket.[12]

Portable consumer electronics such as digital organizers allowed users to create, capture, and document their digital lives. But their storage and design constraints also limited the kinds of information they could create. In time, consumer electronics would predominantly become sites for consuming media and information. The information practices of creating and consuming data resulted in new documents and collections that influenced how people thought of themselves. The early 1990s consumer PC market would come to influence personal identities and notions of taste, value, and style by way of collecting and consuming digital content. For some early PDA users, organizing data with personal computers initially resembled organizing paper files at work. But as more media devices and consumer electronics entered American life—gaming devices, video cameras, and portable cassette players—externalized digital data storage brought a new culture of personal information management involving portable devices and digital files.

These new machines enclosed users' computing environment, limiting their ability to program, tinker with, and contribute to the software they used every day. From the open software and communal computing cultures of the 1970s, personal computing turned into a marketplace of consumer technology targeted to individual users. Early online hobbyist and hacker communities still existed, but a new era of shrink-wrapped software solutions began to emerge. By the early 1990s, software wasn't something that most computer users built and contributed to like graduate students improving and customizing versions of the Unix operating system. Instead, it was something you could buy off the shelf and rely on. The enclosure of software development led to an increased reliance on the extraction of user data as part of the software development process. During this move to software apps as a consumer experience, transmission and wireless networking helped data to take on a new valence of extraction and exchange.

FILES AS REPRESENTATIONS

For more than a century, the idea of the file has persisted, wrapping and structuring information created and managed by institutions. Even in today's cloud computing and shared storage collaboration environments, files and file directories are how we collaborate on documents. Modern filing systems may appear to be like earlier paper systems, which incorporated file architecture for centuries.[13] Think of large steel filing cabinets, manila folders, hanging files with labels on tabs, and drawers that communicate storage structure and access points for retrieval. File directories, like hanging files in filing cabinets, act as a system of orientation, just as they architect storage and delegate resources to memory. Since early operating systems, directories' context and structure have enabled computational data management.[14]

For most users of contemporary systems, files' comforting visual representation supports redundance and backup.[15] However, as a representation of information and data, the file is a powerful signifier; it can both denote an object and not be it at the same time. A file can be both a place and a holder for information. In systems, files assume context but can also carry contextual markers of evidence. Think of files like a painted box in a parking lot that indicates the location of a potential parking space. Because their context is embedded, bundles are easily identifiable by the machine, operating system, and software. This bundling allows data to be easily identified, managed, and retrieved by appropriate applications. Files can be bundled into folders, and bundles of folders can then be sorted into directories. In digital systems, a file directory is a virtual container that groups other files in a structured way, much like a parking lot can order cars on different levels or designate special access. So too can files both represent information and be it, depending on how they are leveraged for access.

A hallmark of files as representations comes from their function as indexical signs within information systems—in other words, their indexical qualities. In *Indexing It All*, Day argues that documents such as files acquire meaning from their origination and in their specific situational use.[16] As such, the indexicality of files means they not only contain data but also carry evidence of contexts in which they were created and used. Such provenance information can create further layers of evidence, in Day's characterization, and provide possibilities for further associations and links. Similarly, Feinberg's theory of metadata emphasizes that documents such as files or databases carry contextual markers or cues that make them meaningful within organizational domains or information retrieval frameworks.[17] Thus, indexicality is what allows files to function in both digital and analog systems by linking the data contained to the context it carries, making it actionable and potentially both retrievable and usable over time.

Another important feature of both digital and paper files is their provisional character; they can be easily amended, extended, and moved. In the last two chapters, we saw how files can be impacted by their format: documents in an individual's dossier, stacks of punch cards, or reels of data tape. It may take time, planning, or human resources to interact with it, but the form of the file persists because it can change while maintaining a logical structure for its containment. Information can be added, updated, moved, and rearranged in files. Such working files can be mobile and easy to retrieve, but they still rely on structure and standards and sometimes dependencies from human or machine readers to make them intelligible and accessible. As the digital file came to replace earlier eras of paper-based information systems, data began to spread because digital files could be duplicated faster. In the process, hierarchical directories with folders became the primary way that users organized their files on PCs. Just as we name files within Dropbox and Google Drive, early users used files and file directories to create working data management workspaces on PCs by naming, renaming, and reordering them. They controlled them first using a command line interface and then later with skeuomorphic icons in graphical user interfaces.

As representations of structured information, the file, folder, and directory came to dominate how people interact, understand, and learn to use personal computers in workplaces. As personal computing increased data creation in the 1980s, PCs and digital handheld devices would incorporate abstractions of files in operating systems, GUIs, and external storage formats. The abstract representation of the file took on new qualities with devices like PDAs and cell phones. New pressure on the file concept from the reality of communication networks expanded early personal telematics to accommodate the fluidity of the file across systems. It enabled data to be organized, stored, transmitted, and synchronized across networks. The indexicality of files thus serves a critical role in organizing and contextualizing information not only within specific domains but across them as well, making the concept of the file both a practical tool and a conceptual heuristic to study individual practices, technology adoption, institutional change, and organizational domains.

The semantics of files and their indexical capacities are incredibly flexible partly due to software design.[18] But ultimately, users may bring perceptions of paper file qualities to data formats that are unrelated to software applications. Because of files' perceived structure and semantic flexibility, their transmission or sharing properties aren't well understood by most users.[19] Although users are in control when organizing their files in directories, a second-order level of syntactic formatting is often needed; to open the file, the correct software application must be launched. Today, mobile operating systems and apps largely interact with files to import information when launched. When we create data with mobile devices, particularly strings of textual information, we assume that it can be easily bundled and moved. Yet, it may be divorced from its original context as it moves across networks, apps, platforms, and devices. Today, mobile devices and apps can confusingly allow users to create many versions or copies of files, storing them in different places, or present media to users as files when instead they are uniform resource locator (URL) links to app features.

In time, data began to move between places at increasing pace and scale and eventually wirelessly across internet-connected networks. The earliest data formats were texts, calls, and emails, but these were quickly overtaken in popularity by documents, spreadsheets, music files, pictures, and games. These new devices relied on the user to organize and update their data, in contrast with the automated recordkeeping found in business, commerce, and government, where organizations and their data managers were subject to laws, processes, and organizational workflows built over decades during the transition from manual data processing to processing machine-readable information at scale. With more digital stuff to interact with, individuals became tasked with managing its integration and storage across dedicated personal devices. Shortly after these personal devices began to connect to wireless networks for data transmission, mobile devices began to subsume files as representations for data with apps.

FROM PORTABLE TO PERSONAL

The first PDAs were the Amstrad PenPad, the Apple Newton MessagePad, and the Casio Zoomer, all released in 1993. Each of these devices was designed with hardware and software principles found in notebook computers from IBM—an evolution from smaller portable machines such as Gateway's Handbook, the Atari Palmbook, the GRiDPAD tablet, and the Fujitsu Pocket Pad. Smaller personal digital organizers from Casio, Sharp, and Hewlett-Packard had similar sizes but could not connect or transmit data from other machines easily. Casio, a Japanese company known for its affordable watches and pocket calculators, wanted to create an affordable, lower-powered PDA. Casio was already a leader in the market with digital diaries and electronic organizers. Partnering with Texas-based computer company Tandy Corporation, they contracted Jeff Hawkins to develop the Zoomer (a play on "consumer") in the late 1980s. Meanwhile, Apple was spending millions developing the Newton MessagePad and its overhyped handwriting-recognition features.

The Zoomer's designer, Jeff Hawkins, left Tandy immediately before its release to start his own company that developed software applications for mobile computing devices,[20] founding Palm Computing in 1992.[21] Hawkins got his start with GRiDPAD tablets, developing handwriting software for stylus input on small screens. He saw an opportunity in developing not only data entry software for portable devices but, more importantly, data transfer. As part of his separation agreement from Tandy, Hawkins was restricted from working on hardware and operating systems that would compete with the Zoomer project, but he still saw a future in customizing features for a dedicated PDA operating system across different device hardware.

Hawkins believed that the future of personal computing would be in smaller, mobile devices capable of quickly receiving data input through gestures. He predicted that PDAs would be everyday tools for professional office workers and high school and college students, replacing their paper day planners and graphing calculators.[22] Unlike Apple's Newton, which

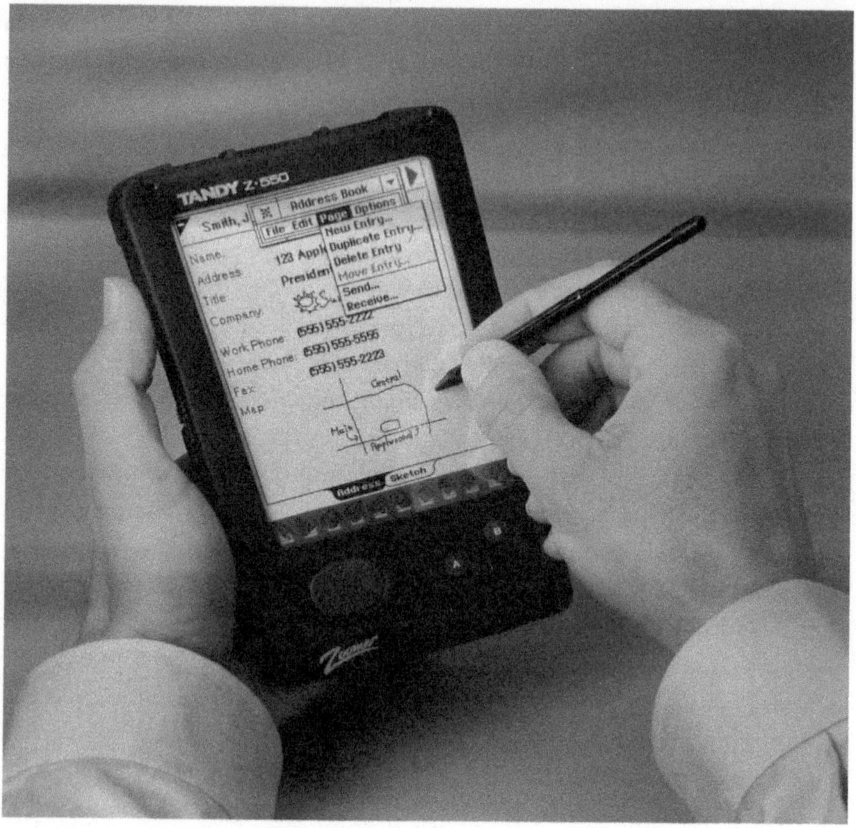

Figure 4.5
The Zoomer PDA from Tandy. Courtesy of the Computer History Museum.

was focusing on software-supported handwriting recognition that would slowly "learn" from users, Hawkins's design approach emphasized rapid data entry. With simple gestures, data could be clipped, copied, and easily moved across PDA apps. PDAs would eventually reflect Hawkins's vision and have powerful operating systems with low-cost, low-power messaging over cellular networks. The future lay in managing data with software that would leverage wireless networks for transmission.

Early reviews of devices such as Hawkins's Zoomer claimed that PDAs were "the first device in a totally new category of computing."[23] Projections of PDAs charted that these new portables would take over this new market: pen-based complements to users' office computers or their home portable computers. While emphasizing their personal information management capabilities, tech journalists predicted that these new portable technologies would improve office management. They would increase employees' productivity by giving them the ability to do more work while traveling and commuting. Their input and retrieval features, as well as the ability to switch across applications, led them to be compared to owning many kinds of information technologies in one: "having a PDA is like having a telephone, fax, PC, pager, and data storage device in your pocket."[24]

Like a personal computer, early PDAs the size of paperback book had a file browser, icons, and the ability to record and store data. These first PDAs were more powerful than digital organizers because they could receive and synchronize data with a user's home computer or work PC with specialized cords—but data synchronization features and data transfer cables were sold separately. Compared to other digital organizers at the time, the stylus pen input feature was lauded as the most exciting feature of these first PDAs. The Zoomer's handwriting recognition was vanguard, but users had to learn the input system. Most histories of PDAs lionize the Newton MessagePad development as a noble attempt to recognize and learn from users' natural handwriting. Apple's first touch-screen PDA was a market failure. But it paved the way for the future of hardware using pressure for screen inputs by incorporating haptics.[25]

This first generation of PDAs sported GUIs with many of the same metaphors, images, and actions for working with files with PC operating systems. But buttons, styluses, and handwriting were how users interacted with the device. Navigation was more than keyboard and mouse clicks; there were often visual elements—both image icons and drop-down menu options to explore features. Storing data that could both live in apps and be easily transportable was the future of PDA software, but the early devices came with preloaded reference information. For example, the Zoomer came with twenty built-in software applications, many with features now common to mobile phones. The PDA had 1 MB of RAM, 640 KB of which was used by the operating system, and could be expanded with flash memory cards. Apps included a world clock, a calculator, and an address book. The Zoomer included more complex icons for a language translator, America Online for news updates, and communication and transmission to send emails or receive faxes. With these new mobile devices and their proto apps, the meaning of the file—as a structured representation for storing information—started to be shaped by the movement and flow of strings (or "snippets") of data across apps.

Early data entry and designs for personal information management made distinctions between information's format and content by way of these strings or "snippets." For example, a calendar appointment might have a name, location, and time. Each of these snippets could be reused in a map, address book, or diary. Strings of data could also be used to infer context of intended use, such as a nine-digit phone number or a five-digit ZIP code. Consider how an email included important metadata in the header that could be used for reporting, timekeeping, and reference. These textual representations in strings of data help us understand the early days of PIM tools. Different kinds of tools, habits, techniques, and feature expectations emerged. Each kind of information associated with a format demanded a different way of interacting with and managing it. In time, software applications embedded these assumptions about strings and snippets so that they could be copied, pasted, and shared across apps. As PDA software began to rely on strings of information to encourage portability, it began to shape data entry as well as representations of structure and containment to support data transmission.

Zoomer executives described the device as having the "power of a PC and the best of pen and paper."[26] But the best affordance of pen and paper is being able to write anything down, fast. For handheld devices to beat out paper day planners and handwriting, users would need to easily navigate across applications and toggle between tasks. For example, they could add an appointment to the day planner while pulling up a contact's location in the address book. However, even with these considerations, processing speed remained a problem with these powerful little devices. It took about fifteen seconds for the Zoomer to turn on and even longer to switch between applications.[27]

Despite its speed issues, the Zoomer allowed users to transfer data to other devices across networks. Much of the transmission was still wired and depended on users connecting PDAs to computers with special cords and ports. The Zoomer had basic file transfer software that used three technologies to connect with different external devices: infrared for exchanging data with other Zoomers, a R5232 social port, and a PCMCIA slot for modems and expanded storage.[28] Early reviews highlighted possible use cases: "Travelers with expense accounts will love Pocket Quicken, a financial manager that can transfer its data to the fullblown desktop version of Quicken"[29] These communication features—and the file compatibility and the interoperability of apps—meant that users could transmit data to other devices across networks.

From the start, there were two ways PDAs supported data creation and access with apps. The first way was with factory-set apps that arrived preloaded with information as part of the device operating system. These relied on local storage, whereas a second kind of app with more dynamic features relied on transmission networks to send data and update apps with new information. Computer reviews of PDAs at the time often mentioned these "preloaded software programs." Unfortunately, early preloaded software applications were not completely integrated across the mobile operating system. For example, to make an appointment with a colleague and bring notes to the lunch meeting, a user would have to go through two separate apps to capture the time and location as well as free text information. But tech journalists predicted that soon integrated software applications would be the future and promise of mobile computing: "when software developers begin releasing their programs in mass distribution the Zoomer's usefulness will increase evermore."[30] The second way relied on networks to access data and are much like the apps we use on cell phones today. Like the software experience using a personal computer, PDA apps like AOL and Pocket Quicken emphasized the potential of network transmission for real-time information updates.[31] It wasn't long before the potential of communication networks could be seen, as these apps and cell phone companies would connect networks to PDAs. By 1994, Bell South cellular had announced the Simon designed by IBM, a PDA that combined a phone, fax, pager, and email device.[32]

Hawkins's Palm Computing (later Palm Inc.) developed apps and a mobile operating system that could capture the burgeoning mobile computer market, which used software called "PalmPrint" as the primary data entry method. Unlike Newton's handwriting recognition program, PalmPrint was basic script for writing letters and numbers. The program used pattern recognition of the script that relied on users to block print letters for input with a stylus. Palm designers and developers tested several design approaches that emphasized the different kinds of user interfaces that should be accounted for with PDAs.[33] Advertisements explained to new users that handwritten entries were not static. They could be moved, indexed, and automatically separated with cut-and-paste features, giving users "all the advantages of computer data" while using a stylus or a handwriting entry.[34] Flexible data input and output enabled the quick reference across PDAs essential for personal information management, and data entry began to shift from files to apps.

Based on experience with Zoomers and the Palm operating system, Palm Computing developed a unified strategy for data entry that could be used across PDA devices. With Power Ink and PalmPrint, the user had the option of inputting script with a stylus or typing on a screen keyboard. Power Ink and PalmPrint were both able to treat textual data as discrete objects or as a structured data format that allowed for data created by users to be moved and cut and pasted across apps on the system. In a widely circulated position paper, the company discussed the three different "kinds" of data that should be managed and supported by PDAs. The first kind of data would be information for personal use and typically stored only on the device. The second kind of data was text that was easily retrieved and copied for later use and other contexts such as copying an address to an email message. Finally, Palm Computing imagined a path for data entry that would allow for information such as a phone number to be spread across different apps and locations in the mobile operating system. By allowing for all three kinds of interactive data entry for users that could be personalized, these methods would lead to broader acceptance from new customers.[35]

KEEPING DATA IN SYNC

PDAs still required users to manage their data by moving it across devices. Manufacturers emphasized tethering PDAs to PCs because they required users to transfer data. But data transfer software and cumbersome cables made it difficult for users to share data between PDAs and their personal computers. The manual transfer process had many steps, and it took time to reconcile versions of files and update information across a user's devices. Ironically, the time that was saved creating and accessing data with a portable device was then lost with slow transfer between machines with connectivity software and proprietary data transfer cords. The chore of this extra coordination drove most PDA users to delay or avoid transferring data altogether.

To address the time-sink problem of transferring data, Palm Computing developed a synchronization process that made it easy for users to synchronize data automatically across machines.[36] Hawkins and the rest of the software engineers at Palm Computing knew that PDA owners didn't want two computers to manage their data in two places. They had collected user data from new PDA owners. When consumers bought Zoomers, they would register their devices and share what they planned to use them for on a registration card. User information and feedback collected by Palm Computing showed that "people didn't necessarily want to own a second computer. They wanted an *accessory* to their PCs; some means of carrying around the data that were also on their hard drives."[37] This value of automated syncing spoke directly to this concern of carrying the most up-to-date information on PDAs and automated synchronization. PalmConnect software could easily synchronize the Zoomer's data and file structure with PC files because it used short strings of text as data objects. Palm's synchronization software linked desktop computer software programs with PDA personal information management apps. PalmConnect would become an operating system for devices and an essential platform for software developers to release custom applications for PDA users.

Synchronization with PalmConnect became popular with both users and PDA software vendors. It became a proof of concept for third-party software development opportunities to transfer and synchronize data from mobile devices.[38] Data entry, combined with bundling and syncing, would create opportunities for other transmission methods for structured data, such as an infrared shortwave transmission method. Synchronizing data across devices also continued to drive the development of the PalmPilot handset series, which dominated the PDA market. Instead of relying on the user to manually transfer data and make backup copies, the device itself could be docked in a cradle connected to a computer, and synchronization would automatically initiate. Syncing using a docking station would allow developers to innovate on the potential of data transmission between users across networks. It also primed users to put more trust in app developers and handsets' capabilities using apps to handle data transmission. For example, PayPal was originally developed as an app in 1999 that used Palm infrared signals to transfer payments between Palm users.

Syncing and other data transmission methods worked around the lack of a robust wireless infrastructure for PDA consumers in the US but often

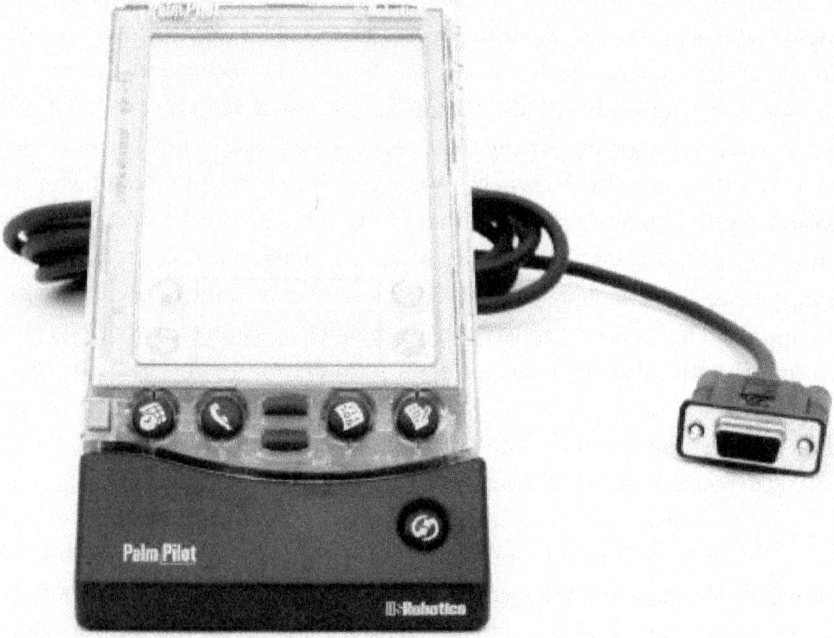

Figure 4.6
PalmPilot in syncing dock station. Courtesy of the Computer History Museum.

seeded the anticipation of wireless connectivity. Many PDA reviews in business news coverage from the early 1990s remarked that accessing data with a device would surely increase productivity for office workers. However, these devices would not penetrate the broader market until "widespread, low cost, wireless data communication becomes available" via mobile cell phone networks.[39] Both digital diaries and personal organizers were designed to create and carry data, turning wherever you were into your office "on the go." But bidirectional wireless communication was still out of reach for these power-hungry devices. For a time, the allure of PDAs such as the Zoomer and the later PalmPilot lay mostly in the ease of automatic data synchronization with PCs.

The significance of PDAs to the early history of mobile computing lay not in the speed of data entry that designers like Hawkins had hoped for. By the early 2000s, "pen computing" and stylus data entry had largely failed. Overall, PDAs' ineffective handwriting recognition and slow response times were why it wasn't more popular with consumers. The PDA stylus and screen input just couldn't compete with the speed of pen and paper in capturing the speed of handwriting and people's ability to quickly recognize handwriting. Instead, the influence of PDAs can be seen in our mobile devices that automatically synchronize data across devices, accounts, and apps. As data transfer between apps and synchronization became more sophisticated, PDAs advanced mobile devices by allowing people to combine most of their personal information management tasks across a few apps. Synchronization was eventually replaced by wireless cellular data transmission that would decompose the representation of files even further. The PIM apps that came as standard features on PDAs, as well as their potential for data synchronization, would reappear in early feature phones. Cellular transmission wouldn't just change individuals' approaches to data management and personal information—it would grow the market for software developers, creating third-party mobile apps that collect and manage data.

Mobile apps handled data transfer and synchronization differently than early PC notebooks and PDAs because developers could count on wireless connections to transmit data more frequently. Apps could seamlessly update and refresh data at the network level and, at the user level, leverage transactional interactions, giving users less control over accessing or organizing file objects. This dual process is how people started to understand the relational affordances of networks supporting data practices. Similar to time-sharing explored in chapter 3, seamless data synchronization relied upon the "constant touch" of wireless networks leveraged through software that took advantage of constant tethering.[40] Truly seamless and faster data synchronization was finally achieved through 2G cellular networks incorporated in a second generation of PDAs in the late 1990s. These "pocket PCs" connected to the internet and mobile networks much like cell phones. Expensive devices, like the Dell Axim and the Hewlett-Packard Pocket, came with PowerPoint and were released in the early 2000s and marketed on exclusive cell phone networks. Many of these second-generation PDAs were targeted to mobile professionals in health care, finance, and sales. Despite the seamless connectivity of second-generation PDAs, they flopped because of high pricing and poor coverage of data broadband necessary to fully make use of pocket PCs.

The early mobile apps built for PDA operating systems such as PayPal payments showed that demand was possible, but wireless technology lacked maturity.[41] Tech companies also found a mismatch between the affordances of mobile devices when compared to PCs. PC capabilities on smaller devices "did nothing but clutter the screen with options that the customer didn't need."[42] Transfer started with synchronizing information to support work and then quickly became personal when people had the ability to communicate across networks. In time, data transmission became widespread and more robust as devices allowed users to harness and connect to wireless cellular network connections. Apps began to offer services that leveraged the capabilities of a constant tethering to a wireless network. So, data creation, and its collection, increased.

FEATURE PHONES AND WIRELESS NETWORKS

Americans were already familiar with wireless networks before PDAs could connect to them.[43] By 1993 when the first PDAs were coming to market, there were approximately sixteen million active cell phone subscriptions and nineteen million pagers connected to networks in the United States.[44] Designs for handsets entered the market, and by the mid-1990s, nearly all of them had similar size, battery life, and portability. They also shared functionality for creating and storing short strings of textual information with data entered manually: an address book, an alarm clock, a calendar, and a messaging client for texting with Short Message Service (SMS). These early handheld devices included the same proto apps to store personal information and alphanumeric strings of data: an address book, a calendar, a notes app, and a text-messaging client. By connecting to wireless networks, cell phones overcame a personal information management problem that PDAs had yet to meet: keeping data synchronized while people moved around.

Early feature phones—cell phones with buttons and built-in software rather than touch screens and user-installed apps—had limited functionality. Users could save small text files within directories such as an address book, like PDAs. But just like early PDAs, there were major hurdles transferring information with cell phones such as text messages, photographs, or even phone numbers from devices. Early feature phones made it easy to send and receive text messages, but limited storage and processing power made it difficult to transfer or save them outside of the device. This lack of transferability was partly due to the assumption that cell phones would be primarily used for voice calls. Cellular networks weren't initially designed to carry packets of data that would be legible for humans to read. Instead, they were designed to carry networking data for machine-driven routing processes. Early cell phones also had limited processing power and internal storage making space for the antenna and battery. However, unlike PDAs that sacrificed speed to expand storage, cell phones had faster processing power and longer battery life.

Cell phone designs privileged networked transmission of data over synchronization initiated by users. Systems typically allowed users to manually delete data to make space. Early generations of feature phones allowed information to be stored to flash memory such as SIM cards and expandable storage slots for flash storage. Over time, smartphones with greater processing power and larger storage capacity replaced early feature phones. Within a decade, smartphones evolved into mobile computers that looked much like what PDAs in the early 1990s had promised but included powerful network transmission capabilities. Eventually, mobile networks would connect to the internet to synchronize and transmit more data—not just calls and texts but also location data, pictures, and video. In the genealogy of mobile handheld devices, feature phones shifted the atomic unit of information contained in files to apps that transmitted and received data automatically when tethered to wireless networks.

Unlike PDAs, structured information stored in cell phones mostly wasn't personal information such as notes or calendar appointments. Instead, it supported the cell phone as a communication device, first for phone calls and eventually SMS messaging and GPS navigation. For example, common phone numbers from contacts were collected in a phone book. The cell phone could allow users to easily speed dial contacts or see the last calls they received. As personal information management extended to mobile devices, data practices were influenced by network connections. There was always the potential for making and receiving calls, as well as storing more numbers and text messages from contacts. Still, these proto apps supported tasks for personal organization of contacts data that would become individualized through personal curation and collection.

With cell phones, personal information practices would become curatorial—choosing what contact information to save or what text messages to delete. Users would manage storage and collect important information from contacts, as well as anticipate a mobile network connection. Individuals would become archons of their own personal digital information when they could create, keep, and share digital information with personal computers. Yet, I argue that the transition to wireless transmission shifted stakes for the management of data over the long term. The emphasis on two-way communication with wireless networks eventually led users to less frequently actively manage personal information. Instead, later apps would capture information about users and funnel it to third parties, further distancing them from their data by way of the constant touch of networks.

If personal computing devices brought data closer to users, letting them manage and synchronize it, mobile devices with limited storage and transmission capabilities pushed them away from their data traces with apps, platforms, and cloud storage architectures. Wireless network connections created a new relationship with data transmission that would subsume the file as a representation and file directory as a structural form of memory to streams of data. This new arrangement of devices, users, software, and network communication infrastructure would result in a loss of control for individual users, as seen in data created in networks built for automated systems to transmit and collect.

The combination of the mobile phone and internet-connected services created a new kind of user. Adrian Mackenzie refers to them as "the network user" who is "a troubled figure, plunging and ascending between freedom and control, sometimes configured as windowless monad (sic), and sometimes a locus that disrupts all closures."[45] With two-way cell communication, networked users were encouraged to collect more information from personal contacts they could communicate with. Once individuals began to use devices that connected to networks, communication and interaction using these technologies required constantly transmitting data for the infrastructure to work—the extraction of user data and metadata in exchange for networked services. This reality created new circuits of control through data exchange. In Wendy Chun's words, "all electronic interactions undermine the control of users by constantly sending involuntary 'representations.'"[46] These representations created conditions for wireless technology and the digital data from human communication supported by networks today. A network user could generate new streams of data valuable to supporting network transactions, but these involuntary representations also normalize extraction, accumulating and aggregating networked data collections of human activity beyond our control.

According to both Mackenzie's and Chun's accounts, wirelessness brings an ascendance of data transmission but declining control over representations created by users. Mackenzie's theory of wirelessness and the network user helps us understand expanding and shifting boundaries of representing data by what mobile devices promised once they were connected to mobile internet services. PDAs like the Zoomer and PalmPilot offered users new attachments to files and synchronization across multiple machines. But smaller, mobile devices connected to wireless networks forced users to favor metadata rather than locally stored data. Such data management designs on mobile devices set the stage for network extraction from third parties.

THE AGE OF APPS

Although millions of more people have owned cell phones then ever owned PDAs released in the 1990s, early apps from these devices presaged how we use our smart phones today.[47] Once mobile phones featured external apps and bundled data, they allowed networked data extraction in automated and less personally managed flows of data funneled through apps. Files become ever more decomposed representations of containment, and transmission is automated. Data is now born in networks and managed by the apps on devices that assume wirelessness. Early cell phones had limited storage capacity and battery life, and service was priced by number of calls and text messages. These resource limits taught users to watch how much memory they had left in their text message inboxes to receive data, as well as storage limits for addresses and media files such as ringtones and photographs. But eventually, personal management of data became less of a priority for users and quickly became a business imperative for app developers. It faded into the background for mobile operating systems and apps as users spent more time interacting and connecting with networked data instead of managing it.

Before the age of apps, there was a longue durée of files. The abstraction of files that we experience today with our phones was imagined in the 1970s, designed and engineered in the 1980s, and entered the mass market in the early 1990s. Hence, files mean different things depending on whether you are an engineer, a bookkeeper, or a graduate student. But files as representations of containment also reveal gaps in our perception of how information is decomposed into strings of data to be processed and handled automatically. Most data about human activity are now created with mobile devices and internet-connected wireless networks, and apps assume synchronization, wireless transmission, and network coverage. With mobile devices and wireless networks, data became controlled through standards, protocols, regimes, and routines of apps that assume network transmission and constant tethering. The abstraction of the file, which influenced the social life of documents, continues to affect our perception of data transmission, storage, and management in peculiar ways.[48] The evolution of files showcases the distancing of storage and the proxy management of personal data. Ultimately, the rise of automated systems and the increased prevalence of software applications assume wireless network coverage. These networked communication experiences led to new sources of data and ways of representing them as metadata, filling corporate data archives documenting uses and users. The age of apps and wireless connection led to a new regime of access. Data is now personal and born networked, ready to be collected by the new archons of memory.

In the age of apps and wireless networks, data is always accessible by third parties such as software developers, mobile service providers, and handset manufacturers. As cell phone functionality broadened, users gained the ability to capture formats with larger data footprints, such as photographs and audio. More space had to be found by off-loading storage data to platform services and cloud storage services. By the time smartphones arrived and next-generation networks connected them with faster processing, many mobile operating systems began automating the "digital housekeeping" that PIM tools delegated to users of early PDAs and PCs. As data generation and data collection increased through cell phones, data transmission for network users became automated, dangerously distancing users from an awareness of data.

Later, as data extraction and storage became coordinated and automated, over-the-air updates, automatic backup, and cloud storage became standard smartphone features by the 2000s. With apps and cloud backup at the center of mobile computing, stewardship and PIM practices moved from users' purview to apps developed by other third parties and the stacks of infrastructure that supported apps. Like PDA synchronization software, early mobile apps tried to reduce users' effort and shift their attention away from management to interaction. However, organizing and transferring data via third-party apps and platform services created new sites of extraction while culling access points and storage possibilities for local data. By 2007, around the release of the iPhone, the interface design of apps for smartphones and mobile operating systems began to hide file directories—which had been important to filing strategies of earlier computing practices since the 1980s—from users.

As mobile apps and operating systems began to increase data collection and automate data management for users, cloud storage for unlimited user-generated content became part of using not only a smartphone but apps and platforms as well. Alongside this change, data became relational, and new opportunities for automated networked data collection arrived, becoming infrastructurally useful and economically beneficial for third parties such as software developers, consumer data brokers, and state surveillance efforts. With less personal data management came increased tacit and accelerated data extraction from networks by third parties such as platforms. New data management utilities appear to be bottomless with next-generation networks and cloud storage architecture. But these new automated syncing opportunities came at the price of long-term storage at a distance—what Cathy Marshall has called the "radical ephemeralism" of personal information created with platform-driven digital devices.[49]

Aggregating data from individuals and assembling data archives is more than a technical story about externalization and storage, but it closely tied with the how information is represented, whether in files, data, or apps. For many private corporate bureaucracies, assembling caches of data for prediction had been part of their bid for legitimacy and profitability since the 1970s. The National Data Center proposal did garner the first congressional attention and public awareness about the right to privacy resulting from the collection of personal information from American citizens. However, ensuing debate and legislation only erected safeguards from the government collecting personal data from citizens. But private corporations were not included in the 1974 Privacy Act, nor were they included in the other legislation governing federal data banks. The concerns that alarmists had about connecting and matching information from people's personal files was a reality for many American corporations by the late 1980s.[50] As a result, private bureaucratic organizations such as insurance companies and credit card companies were free to quietly build up data banks of Americans' information and personal data during this transition from digital storage to automated recordkeeping and real-time databases.

Once wireless users began to harness the relational power of telecommunication networks, personal information management became a prime target for the innovation of companies collecting data at scale. Smaller, mobile devices such as PDAs and cell phones created new contexts for users to leverage wireless networks for communication and data transmission. Once data began to be stored in platforms and cloud computing architectures, information was further decomposed to strings and packets of data, no longer resembling structured information formats such as a document or a spreadsheet. Users no longer needed to act as their own personal information manager, steward, or data archivist. In time, the grammar of action once embedded in folders, file directories, and structured information formats gave way to apps syncing user data across networks. In short order, cell phone owners would become users with many accounts, storing and managing data via cloud architecture governed with end-user license agreements or platform terms of service (TOS).

The radical ephemeralism of early personal digital assistants and mobile devices opened space for personalization and synchronization of data. New software solutions for automated recordkeeping and data management aimed to assist users automate and control the flood of digital information they created. Simultaneously, data synchronization resulted in market opportunities and network logics where constant collection, unlimited backup, and distributed storage created new forms of power and control over data from mobile devices. Once users agreed to provide their data in exchange for access to networked services, an irreversible course was charted toward profiting from extraction, automating data management to assemble big data accessible to many different stakeholders.

5

MAKING DATA WITH A CELL TOWER

On Sundays, my phone generates a "screen time" report for the past week of activity. My screen time on my cell phone this week is down 26 percent from the week before when I was traveling. The activity report says that I spent approximately one hour and twenty-nine minutes each day looking at my screen while using nineteen different apps this week. Usually, my favorite apps allow me to listen to music and podcasts, but I probably use my work email and navigation apps the most. By reading my latest screen-time report and comparing it to the week when I was traveling, it appears I used my phone more for text messaging and taking photos on the road than when I'm at home or teaching on campus.

While I was traveling and used different apps more than usual, my phone was constantly connecting to cell towers around me. As I move, I cross patches in a cellular network. Each patch is supported by a few towers that gather and transmit signals via packets of data. This process of tethering to a cellular network is hidden from me, but running in the background while my phone is on, it is constantly sending and receiving signals as I moved through different "cells" of space.

I'm a T-Mobile customer. So, as I was traveling across state lines, a unique SIM card allowed my phone to connect to different towers that make up a quilt of cells for the network's coverage. These towers, and their base stations that store and forward information, automatically coordinate my precise location to carry my messages, emails, and phone calls back and forth across wireless networks and other internet-connected infrastructure. Because of all this coordination, when my phone is connecting to different cell towers for coverage, layers of networking data are acquired during routing for transmission. These layers of data about transmission (or metadata) that are supporting what I'm doing with my phone are automatically saved by some layer in this stack of wireless infrastructure that is made up of cell towers, base stations, and phones we carry.

"The stack" is a layered network architecture made up of hardware, firmware, protocols, mobile operating systems, and apps. Each layer interacts to relay cellular network information from mobile devices and activity within cells. This complex choreography is easily forgotten, and depending on context, the data gathered from it can take on varying degrees of significance. Some of this information that is collected about my actions through my cell phone I can make sense of and is meaningful to me in the moment or upon reflection after a few days—for example, to understand my screen-time habits or my travel patterns. But much of the data and metadata that results from using my cell phone connected to a part of the stack that is inherent (and essential) to wireless communication would be meaningless to me after a short period of time.

Experiencing wireless infrastructure like mobile networks and using cell phones without pause points to another significant step in the evolution of archiving data today—that is, the massive amounts of automated metadata created as part of our everyday human activities that are now constantly collected and then redistributed in different contexts of use. And cell towers are a peculiar type of archiving machine that creates both closeness and distance between data and people that enables this shift in data archiving. In part, this is because much of the big data that we create today is enabled by stacks of wireless infrastructure and internet-connected communication technologies. Much of the big data that is enabled by infrastructure such as cell towers is quite literally not accessible to us because it is masked during transmission or prioritized for automated processes solely handled by machines. But even if we have little intentional engagement with most of the data we make when using cell towers and cell phones, data transmitted through telecommunication networks is many things at once to other stakeholders. Once aggregated and assembled as collections of big data, telecommunications metadata is useful for mobile service providers like T-Mobile or Verizon for short periods to support billing and measuring network capabilities. But during events series and over long periods of time, these same data collected from cell phones and cell phone towers that are aggregated from groups of people can be used for emergency services, public health efforts such as contact tracing, or state surveillance of student protests.

Today, platforms and most organizations demand "big data" from human activities to operate, and this imperative to collect data or tap into existing sources for potential analytics applications drives the expansion of archiving efforts to reuse data collected from networks and sensors. More and more, this comes from our phones and mobile app ecosystems tethered to the stack generating big data from human activities and enabling its collection and reuse. As we have seen in earlier chapters, the promise of aggregating data in digital formats has powerful social and legal consequences, but larger scales of fine-grained information from wireless infrastructure increase the prospective value of high-volume sources of information such as cell phone records and mobile activity data. Tracking human activities online, increasingly with mobile devices, and the resulting big data archives are foundational to the business of platform expansion for companies like Google, Meta, and Amazon.

Metadata from user activities in information systems has a long history in traditional collection environments such as research libraries, but using metadata as evidence of prediction and building profiles only became possible with data management techniques and information processing for digital data formats.[1] Since the late 1960s, collecting transactional metadata from Americans' purchasing habits and media consumption to assemble data archives from consumers has been industry practice in insurance, banking, advertising, and telecommunications. But unlike the public backlash to federal data collection covered in chapter 2, corporate data collection has little government oversite in the recent history of the US.

Until recently, there has been no national legal framework in the US that regulated the massive collection of user data beyond the contractual notice and consent to terms of service.[2] Instead of protecting a user's data through privacy, personal data protection in the US has been left to industry self-regulation, largely because users' typically exchange some or all their data in exchange for many information and communication services.

During the Obama administration (2009–2017), the commercialization of online personal data thrived, growing the platforms that dominate today and creating prime big data sources to collect and redistribute.[3] Once data has been collected by a service provider in bulk, it can be repurposed and transformed and then stored for future uses. With processes such as cleaning, normalization, and chaining, different streams of data can be linked together, resulting in significant transformations. With transformation of data into new aggregated resources, the current terms of service and legal frameworks in the US allow data caches, once collected, to be to be stored indefinitely.

Earlier chapters introduced the technical and social conditions for data to become mobile, mutable, and personal. The previous chapter set the stage to see how personal data would be moved and managed, with networked infrastructure relying on time-sharing utilities and resulting in massive collections of data that needed both management and storage once wireless networks, the internet, and personal computing took hold. User generated metadata from the stack is abundant, necessary, and constantly captured in networks. So, its collection, reuse, and redistribution to feed big data archives help us understand how managing data at greater scales shapes the kinds of evidence that can be marshalled from preserving it. I argue that this constant collection of personal data used for repurposing into big data archives is an epistemic mark of our contemporary moment.

But the management of these "big" data archives at scale can cut both ways for generating reliable evidence and long-term value for stakeholders. Because big data collection is, by and large, wholly automated by leveraging layers of the stack in successive relays, it is instructive to see how value and meaning are assessed, trusted, and tested as data moves through the stack to successive collection applications. In this chapter, I'll argue that big data archiving at scale can create limitations on applications and future use in unexpected ways. These limits can prove to be corrosive to evidence. To explore the limits of big data archiving, this chapter focuses on the expanded frame of data and metadata collection that make up data archives collected by the NSA from telecommunications providers, specifically the call details records collected from the agency from 2015 to 2018 (hereafter "the CDR program"). The failure of this government data archive shows how its value and predictive potential as big data were largely misplaced.

The establishment of the CDR collection program, its operation, lack of efficacy, and the decision to end it encapsulate how the valuation of data and metadata shifted once they were used beyond their original contexts of collection and reused for signals intelligence. The value of big data archives over time may increase or decrease based on their use, despite their size and interoperability with other systems. Indeed, their large volume and interoperability may promise more evidence. But it paradoxically may contribute to their devaluation, diminishment as evidence, and expiration over time.

Often, when big data archives are amassed, their evidential value is assumed at face value because of scale alone: "big" data is often assumed to have inherently more meaning with evidence yet unrealized, lying dormant.

But when big data archives fail, like the CDR program, the lines between use, value, and the production of evidence come to the foreground. Despite networked wireless infrastructures that increase the availability of data that can be collected and connected, these bulk data resources do not always fulfill the possibilities of actionable intelligence or predictive analytics, and assumptions about collecting at scales that promise reliable and trustworthy information are tested.

DESTROYING A BIG DATA ARCHIVE

In October 2018, the National Security Agency shut down the CDR program and purged all the program's data. In a public announcement, the agency reported the destruction of the data it had assembled: "because it was infeasible to identify and isolate properly produced data, the NSA concluded that it should not use any of the CDRs" stored in the archive.[4] From its establishment in 2015 until the program ended, hundreds of millions of metadata records were collected by the intelligence community to thwart terrorism. The CDR program and its predecessor are some of the largest data archives ever to be assembled by the US government, containing millions of records documenting Americans' communication patterns using mobile and telephone networks.

The Privacy and Civil Liberties Oversight Board investigated the NSA's decision to suspend the CDR program and destroy the telephony data, finding that most information generated from the data collection was redundant. The investigation also found that much of the information collected by the program already existed in previous NSA archives and other intelligence efforts. Out of millions of records gathered from phone activity throughout the country, only two reports had resulted in new information from the program's archive over its four years of operation, and only one report "provided unique information about a telephone number, previously known to US authorities."[5] In addition to massive redundancies swept up in duplicated data, there were concerns the program's collection process was ingesting bad data from telecommunications providers, including sources that the agency could not legally use.

Ultimately, the value of this massive data archiving effort was undermined by its lack of efficacy and the possible legal missteps the automated pipeline produced in gathering too much data. The bigger the collection became, the faster it became redundant, and its usefulness to security officials would expire. The larger scales of metadata accumulation appeared to multiply the noisy data the system took in, thus complicating both the meaning and evidence that could be inferred from these call records as activities from individuals and connections to their known associates.

The decision to end the CDR program and destroy its data archives reversed the government's stance on the significance of collecting telecommunications metadata for surveillance and defense against terrorism since 9/11 and earlier bulk signals intelligence approaches from the Cold War era. The surveillance program's problems with isolating reliable evidence reveals much about how data—once collected and reassembled into a big data archive—can be repurposed in contexts beyond data's origination as well as across stacks. Although the format of telephony metadata enabled the linking and assembling of massive collections, it may have resulted in introducing more complexities and error at higher rates.

As we have seen in chapters 3 and 4, aggregating data from secondary contexts has a long history in migrating from one format to another, but with big data sources that can be assembled, sliced, and chained for redistribution, new uses can appear like nails in search of a hammer. Much of the potentialities of reuse rely on how context is standardized and represented in systems—how data is described, formatted, and represented using metadata standards. The ways that data sources are described via metadata formats take on inherent benefits of secondary use by virtue of its ability to link concepts and integrate sources.

Metadata is essential for refashioning and applying data resources to contexts beyond their original domain, such as signals intelligence and surveillance. Metadata is also necessary for data interchange and interoperability across systems. Reliable and standardized metadata is where potential secondary use (or reuse) can be leveraged for most data analysis at scale. While the collection of communications metadata for signals intelligence (that threatened privacy rights) were well known to the American public and had been debated since the 1970s, this data archiving failure at the NSA was met with little fanfare when it was disclosed. Instead of being a boon for privacy advocates, the data archive's destruction was quietly announced and characterized as a cost-saving measure in the face of government overspending. The decision to shut down the program reveals the costs of data collection and linking massive, population-level data sources can be overestimated when the limits of meaningful reuse are tested.

Earlier that summer in June, the NSA announced that it had deleted three years of records and stopped data collection due to "technical irregularities" from the data provided by telecommunications providers.[6] From 2016 to 2017, the collection of CDRs had ballooned 250 percent, but with fewer outcomes and less actionable intelligence.[7] The shutdown was justified by the agency because the aggregated data lacked utility, not over public objections about the massive scope of the collection or potential privacy risks.[8] In the process of amassing a big data archive of telecommunications metadata, the archive became ineffectual in less than five years. It rapidly filled with noisy data that wasn't useful for the intelligence community to identify and predict actions of suspected terrorists. In other words, the CDR program became "too big to know." But how could such a colossal data archive become so useless so quickly?

ORIGINS OF THE CDR PROGRAM: COLLECTING CONTEXT, NOT CONTENT

The CDR program emerged from the 2015 USA Freedom Act, which aimed to curb the mass collection of Americans' telecommunication data by limiting how the intelligence community could collect data from telephone service providers.[9] The program was designed to replace a previous program that collected telephony metadata in bulk from American citizens under the 2001 USA Patriot Act. Public outcry after documents leaked by Edward Snowden revealed that a Foreign Intelligence Surveillance Court (FISC) required Verizon to provide the NSA with daily CDRs (among other disclosures).[10] Congress responded to the Snowden disclosures by ending warrantless bulk metadata collection of Americans' communications under section 215, creating a new design under the USA Freedom Act for metadata collection from telecoms and internet service providers.

Because they are generated as part of network telecommunication services, CDRs are used as both communications metadata for services and business records for billing. As such they are useful for varied amounts of time depending on context and resolution, such as measuring network traffic in a rural area or during a dense sporting event. The USA Freedom Act requires the government to seek a FISC order before requesting CDRs to ensure that surveillance is targeted and limited in scope to two "hops," placing turnover of CDRs to the NSA into the hands of telecommunication providers because they use CDRs in providing cellular network services.[11]

Limiting CDR capture to two hops still results in a massive amount of signals intelligence data from an individual and the network of people they contact using mobile communication. A "hop" is defined as an additional target who is a known contact of the initial surveillance target. The first hop captures all CDRs generated by the initial target's communication activities. The second hop extends this capture to all CDRs created by the contacts of the initial target, thereby significantly expanding the data collected.[12] Although the USA Freedom Act aimed to reduce the scope of data access by eliminating warrantless bulk collection, the two-hop method still produces extensive amounts of metadata and data to assess. The 2015 CDR program placed the responsibility for responding to FISC requests and sifting to the NSA on the telecommunication companies, which, as discussed below, can introduce possible errors and even more data.

It is instructive to step back and consider how CDRs are generated as part of telecommunications services and compare how these data can be used as signals intelligence in surveillance. Call detail records are made up of data generated from actions of people and machines connected via telecommunication networks—typically a phone call or a text message. In addition to identifying the details on how a session begins and ends, they also reveal technical processes of automated network handling and data transmission.

To understand why the collection of CDR metadata for surveillance has been justified, we need to know what data was collected from call records and what kinds of evidence they became once metadata evolved from telephone networks to internet-connected mobile networks. Simply put, CDRs capture details about calls (or "sessions") between one device and another. The details bundled up in CDRs include session-identifying information (including an originating or terminating telephone number, an international mobile subscriber identity number, or an international mobile station equipment identity [IMEI] number), a telephone calling card number, the time and length of a call, and whether the call was initiated from a hardwired telephone or a mobile phone.

CDR records are metadata—a unique type of structured data describing streams of individual actions and machine processes. In other words, they capture information from an assemblage of humans and machines in a network. But when collected and assembled, CDRs become data as well.[13] When connected and stitched together, CDRs create a rich, contextual snapshot of thousands of human activities among individuals, small groups, and large social networks. In addition to the details captured above, CDRs also contain the telecommunications route by which the call was initiated and connected. For store-and-forward networks made up of cell towers and base stations that support wireless communications, metadata from CDRs reveal where the transmission started, if the handset was moving, how fast it moved, and where it traveled. As essential components of the wireless stack, CDRs are metadata for telecommunication networking, but the significance of metadata such as location data within a cell over time takes on different meanings as products or processes, depending on how it is used and applied, as in cases such as surveillance capture or human activity tracking.

The power to name network transactions with metadata for both humans and machines has unanticipated, powerful effects when data is collected longitudinally and at scale.[14] In real time, it can be used for routing incoming calls or text messages from cells towers to mobile devices. Without metadata from CDRs generated and collected by service providers, the wireless network connectivity that supports telecommunications networks would not be possible.[15] Telecommunication networks use store-and-forward architecture to enroll CDR data in different ways to hand off information to towers as well as other providers. With the rise of telecommunication networks that connect to the internet over IPs, the significance of these transmission records has expanded as well.[16] As telecommunication networking operators leverage CDR metadata to do their work of connecting ICTs, they support users making calls and sending text messages with cell towers and other telecommunications infrastructure. But as business records, CDRs also allow companies to use the same information for billing and transmission purposes. When used for prediction and tracking, aggregated CDRs can forecast routes people might take based on previous behaviors such as traffic patterns or frequent text messaging with contacts.

Importantly, because this metadata is used by companies to operate their business services, the US Supreme Court maintains that information gathered from CDRs is not protected under the Fourth Amendment. Under the third-party doctrine, people who share information with service providers do not have reasonable expectations to privacy. So, routing metadata gathered by telecommunication companies, internet service providers, and other networking technologies can be collected and reused for any number of applications. Much of the contextual data gathered in CDRs is automatically generated by the networks for routing, which acquires traces of data as the transmission occurs between mobile devices and cell towers.

All telecommunication service providers incorporate these data transmission standards to establish handoffs and to measure the health of network coverage services. Such measures include data about transmission handoffs between network providers, such as Verizon to T-Mobile, or across borders from a US-based provider to another in Brazil or India. They include information about when a phone call starts and ends, how it ends, and if the call is dropped, for example. Further, the data that is collected has been standardized by international formats that are governed both by standards engineering bodies and nation-states such as GSM or UMTS.

In addition to the type of transmission data, the route through which the transmission handling occurs across cells is also collected. This includes handoff receipts from base stations receiving and transmitting data across the network. So, CDRs also contain detailed time information, such as the duration combined with place. Such location and duration data allows CDRs to be used to measure actions or movement across geographic areas. For example, companies can tell how fast a person is moving (e.g., in a bus or on a bike) or how many other people with phones are nearby connecting to the same cell area. Further information specific to devices and SIM cards is also included in the records. When unique SIM cards and IMEI numbers are combined with times and location of transmission paths, a picture comes into view that gives visibility to people's environments and past actions. When assembled, a high resolution of precise activities can be inferred.

For telecommunications providers, CDRs are essential to providing network services, measuring the traffic flows for network administration, and billing customers. But the information that can be made from CDRs is also valuable for knowing broader patterns of human activity. Collections of mobile data patterns are incredibly valuable for data-driven scientific inquiry, particularly among computational social scientists and public health researchers.[17] Even over brief periods of time, data collected from CDRs can be used to infer traffic patterns, the spread of disease,[18] financial fraud,[19] weather events, and disaster response.[20] In part, these patterns can be applied to many phenomena because CDR metadata are standardized for transmission, and cell networks cover large portions of places that people occupy. Further, the frequency with which this sensor data from wireless networks is generated (nearly every few seconds) creates massive volumes of data sources with a particularized granularity of action.

Historically, the legal justification for bulk collection of communications metadata has fallen under law enforcement and state security legislation such as the USA Patriot Act. But unlike wiretapping or monitoring calls, metadata records do not contain the *content* of calls, only their context. When defending the bulk collection of telephony metadata from millions of American citizens after the 2013 Snowden leaks, President Obama famously said, "When it comes to telephone calls, nobody is listening to your telephone calls. That's not what this program is about. As was indicated, what the intelligence community is doing is looking at phone numbers and durations of calls. They are not looking at people's names, and they're not looking at content."[21]

The distinction between content and context when defining the data that CDRs carry is often debated because, with enough contextual information, content can be easily inferred. Depending on its application, aggregated metadata can stand as actionable data for different use cases. For example, General Counsel Stewart Baker later explained, "Metadata absolutely tells you everything about somebody's life. If you have enough metadata, you don't really need content."[22]

Importantly, whether building knowledge in a surveillance context or knowledge about public health or emergency services, the more metadata that has been collected, the more content can be inferred. The inferential value of CDRs has to do with its ontology as a format—they are lightweight for transmission, precise for location routing, personally identifiable for billing and service exchange, and interoperable across many different layers of the stack.[23] Because metadata from mobile networks contains information about time and place of all devices using services, when it is combined, it can become aggregated user data about all the people collectively using network services. So, for surveillance intelligence or measuring population health, aggregated information that contributes to context can create a more precise vantage point for generating knowledge about communities or groups of people than the data from content alone. With high rates of mobile use throughout the world and across all age brackets and because of the format of CDRs, population-level datasets are being generated and collected constantly from the metadata generated with networks.

An overall argument of this book is that archiving data through networked technologies such as wireless infrastructure has changed our culture of memory and the ways we produce knowledge. Telephony metadata is one exemplar source of the big data archives being assembled and redistributed today from stacks of mobile networks. Once data can be externalized and reliably stored in standardized formats, it can be managed at increasing scales because of its mobility. But mutable and mobile data formats that can easily be reassembled do not ensure stable meaning or evidence as data moves across contexts origination to different applications and reuse.

This chapter unpacks the significance of particulation and appraisal processes to big data archiving to understand the (often faulty) reasoning of automatic data collection without valuation. There is an assumption that data that is collected from the stack and then repurposed will both hold its initial evidential value and take on new significance in different applications. But such massive and frequent automated data sources may be at risk of drift, accuracy, or expiration—what critical data studies scholars have described as "the fallacy of big data empiricism."[24] Several mistaken beliefs drive the cycle of big data empiricism that is used to justify its constant collection and archiving. The dominant perspective is that big data can provide a full and accurate resolution of the world. In turn, swaths of data can speak for themselves, transcending context- or domain-specific knowledge practices.[25] But these misconceptions often presume no distinction between data and metadata and rely on partial description for representation. However, the interoperability of metadata for easy redistribution and secondary use can be both a poison and a cure.

Big data enablers, like the stack of wireless infrastructure, rely on easily exchangeable metadata formats, and automated processes to support transmission can also contribute to evidential shortcomings with cascading effects. Although it may support the quick accumulation and storage of data for easy reuse, the resulting big data archives may become "too big to know" beyond their original context of creation, causing issues for assessing authenticity, and quality, or risk overfitting.

The CDR program is an example where big data archiving without appraisal went awry. Appraisal is a concept from archival theory that relies on the long-term valuation. Far from estimating an initial cost, archival appraisal considers the value of reuse and redistribution of information objects by determining origin and trustworthiness. Today, appraising big data archives such as the CDR program relies not on the status of its content but rather on its originating context, frequency, and scale.

THE POWER OF METADATA

Because CDRs are at once both business records and personal data, they are heavily standardized metadata but less regulated and protected than personal information.[26] The context versus content debate between surveillance critics and proponents of bulk-collecting communications metadata reveals much about the importance of assessing the value of big data to know the past and to trust what it reveals or to predict the future. The line between data and metadata may be unclear, but data archives, like any other information system, can't work without it.[27] Although communications metadata under US law has sometimes focused on individual privacy with telecommunications, the contract adequacy between consumers and service providers is typically prioritized with IP-connected networks. Indeed, most terms of service for internet communications today force users in the US to give away their activity data as a condition of using cell phones, search engines, social networks, and platforms. This open regulatory approach diminishes expectations to personal data privacy because of the broad realities third parties such as platforms, app, services, and data handling of user data for commerce and networked services.

The lack of personal data protection regulation that has accounted for the rise in massive data caches of consumer data belies the power of leveraging metadata for building big data archives.[28] But less is known about how data is transformed once it is stitched and linked together with metadata, as in the case of the Freedom Act, which aimed to cut back on bulk collection by limiting surveillance targeting to gathering two hops of CDRs, or in the 2016 Cambridge Analytica scandal, where Facebook "friend of friend" metadata was used to build data profiles of expansive friend networks for political targeting.[29]

Much of our current personal data protection regime fails to grapple with the relational way that data born in networks is stored, analyzed, and utilized beyond simply making calls and receiving text messages.[30] How value is ascribed to collections made of data and metadata that is born in networks cannot always be predicted, but it is nearly always anticipated that value will increase as data is accrued. Metadata are often described as compounding interest—the potential value of big data collections grow and age is considered allative, bringing power to the metadata used to represent them.[31]

Metadata from personal telecommunications, when collected in bulk, can become a valuable stand-alone data resource for determining past patterns of action and predicting future patterns not only of individuals but of groups of people as well. Thus, much the same way that platforms collect user data to target advertisements, telemetry data and communications metadata can be used to create profiles of people for surveillance and counterterrorism efforts. But unlike the content of communications, US legal frameworks dating back to the late 1960s treat metadata from communications technologies—dialing, routing, addressing, and signaling—differently than the content of calls, emails, or text messages.[32]

Protecting communications metadata by characterizing it as personal data is difficult because users have already consented to third parties accessing metadata to transmit content. Like an envelope sent in the post, the information wrapping the message that supports routing and identification is not considered private. But unlike hand-delivered mail, nearly all communications metadata is automatically processed by machines, and with automated transmission, its data management expands to include metadata about these processes and the knowledge that can be gained from representing these processes accurately. Thus, metadata allows for information infrastructures to "act indexically" as they transmit, join, and assemble collections of data.[33]

Information scholars have theorized the power of representing data within systems and how such standardized representations impact the material terms of evidential transmission. For information scientists such as Christine Borgman, the distinction between data and metadata is in the eye of the beholder. This is especially true when examining scientific data practices where different expertise leverages data and metadata differently,[34] whereas other scholars concerned with information retrieval, such as Michael Buckland, regard metadata not as information objects but rather as infrastructure—a mark of its functional administration and use with systems.[35]

However, much of today's infrastructural stack connects users more intensively to providers because the inherent exchange of data needed for networking relies on automated metadata generation. For some users, the metadata that they create when using everyday ICTs like mobile phones or email clients may be elided, invisible, or inconsequential and without much intention. Alternatively, for activists, saboteurs, or hackers seeking to influence algorithmic systems, metadata can be manipulated, masked, or highjacked to circumvent platform controls such as content moderation or government censorship.[36] For data managers or collections administrators, automated metadata can provide insight into systems operations, supporting workflows and organizational processes. As such, the "meaning" of metadata or its significance when assessed can change, depending on the intended goals and applications.

Because metadata supports both evidence and accountability for internal and external uses beyond the contexts of data and metadata origination, some data scholars have suggested leveraging these spectral and shifting qualities. Matthew Mayernik argues that metadata processes are on a spectrum between the practical accountabilities of an organization's goals and external accountabilities to those stakeholders who may access and rely upon the data resources generated.[37] In the case of the CDR metadata, meaning can be drawn from apparent use, but its meaning or evidential significance may become diluted the more metadata is aggregated and used as signals intelligence data.

Even though it aimed to curb bulk data collection from the Patriot Act's CDR program, the Freedom Act's CDR program faced a fatal flaw in gathering metadata from the stack and then using it as data for intelligence. Once collected, CDRs without content could not be sufficiently "particulated"—that is, broken down again into smaller units but for different means from its original purposes. The stack architecture of networked infrastructure makes it difficult to collect data and particulate it appropriately without collecting it in bulk and gathering noisy or bad data at the same time.[38] Even today, despite aiming to curb collection overreach, the CDR program still allows for massive amounts of personal information to be collected, even as it risks NSA not being able to use it legally if they receive too much from providers. With the Freedom Act's CDR program, telecommunication providers compile their business records and turn over CDRs to the NSA based on government-provided queries. Providers are not just passive in this process; they comply with records requests and data collection at great expense.[39]

ARCHIVING BIG DATA

Signals intelligence surveillance such as the CDR Program under the USA Freedom Act assumes sourcing from networked telecommunications metadata. Such metadata is indelibly shaped by the business logic that user data can be freely collected and repurposed for applications and contexts beyond communications services. By gathering and stitching together massive amounts of telecommunications metadata, these new sources of big data have been put to service in the name of prediction to become big data that can be leveraged beyond commerce to state governance. These big data archiving projects rely on the architecture of telecommunications, wireless network services, and internet-connected platforms by extending uses for data from the stack to accumulation. So, the ideology of constant data collection presents itself as an empirical fact—a technological necessity for network transmission and networked communication. But this assumption is not without limits for archiving processes and the knowledge projects that depend on them.

Both copper-wire telephone networks and wireless network architecture require (and generate) more metadata to transmit packets of digital information through stacks of infrastructure. The increasing pace and layers of telecommunications metadata needed to support everyday ICTs draw our attention to the machine content of these transactions' relationality, such as the many cell towers that hand off messages as I am traveling through a network. The complexities of digital communications acquire layers of metadata as part of establishing transmission and ensuring reception across networks. Metadata links of in-betweenness bear many relations across contexts, vectors, or frames of creation. As more information is transmitted across networks, the more telecommunications and IP providers generate and collect more communications metadata from populations. Not only is storing such data a part of the infrastructure of wirelessness—managing it becomes a part of service provider's business operations. And as digital communication networks shifted to internet protocols, the intelligence communities started to push for more access to metadata because content is now increasingly encrypted.[40]

As wireless networks increase layers of metadata during handling, the significance of collection and indexicality come into play. The intelligence community's ability to collect metadata, even when decreased and limited to two hops of data collection, resulted in massive amounts of extraneous records, including noisy data such as spam calls, telemarketing calls, and automated "robo" calls. Both bulk collection and targeted metadata collection potential for particulation are not just legal matters, as these processes face technical limits too. In exploring the benefits and constraints of signals intelligence from bulk metadata collection, a National Research Council committee concluded that there "was no technical substitute for some aspects of bulk collection."[41]

Collections of metadata from telecommunications networks become haystacks of big data quickly. These population-level data collections rely on legal regimes that diminish users' privacy, increase advertising, and offer insufficient personal data protections. Such haystacks produce an overly complex and opaque access regime in which data is always assumed to be collected for its future value.[42] But putting faith in haystacks as means of foolproof prediction neglects realities of how such resources might be misused or fall short in the big data era. Such naïve appraisal relies on a future built upon narratives of prediction and speculation, underwritten with immense environmental and economic costs of seemingly unlimited storage and rapid retrieval capabilities.

If unbridled data collection made up of network signals is an epistemic mark of big data archiving, then speculation about its future access value and reuse possibilities is part of the dataist beliefs of prediction. Value placed on the speculative possibility for unlimited data reuse then inhibits appraisal and curation at the time of collection and relies on massive resources to ensure data management and preservation. Further, storage constraints of data at scale outweigh the contexts and value of particulation. Traditional archives as infrastructures grapple with this drift as a condition of their operation but account for this by relying on principles of appraisal and valuation before initiating collection in ways that big data archiving projects do not.

Indeed, much of the monetary investment in big data archiving projects results in data resources that are presumed to have economic value as actionable evidence. However, the evidential, functional, and informational value that is ascribed to big data is often pressure tested after it is gathered into collections. For archivists and preservationists in institutional contexts, appraisal is the process of assessing the archival value of collecting materials before massive projects commence as well. It isn't just about weighing the cost of preserving a collection because of the unique information it contains. The appraisal process also considers the functional, evidential, and informational qualities that may inform a collection's enduring value.

Like the intelligence community's practice of bulk metadata collection, archival appraisal relies on estimating future value, but the enduring value of big data can be overestimated. Much of the data archives assembled from telephony metadata and records born in networks are built upon narratives of prediction and speculation underwritten with seemingly unlimited storage and assumptions about instant search. Gathering data in bulk then becomes the imprimatur of haystack models of knowing and building knowledge—quite different from traditional archives, in which the destruction of records is a routine part of the appraisal process. The capacity of unlimited storage and search are further limited by access regimes that are increasingly governed by platforms or subject to platform models of access.

LIMITS TO HAYSTACK WAYS OF KNOWING

Big data archives that fail, such as the CDR project, reveal why platforms and cloud storage services lack traditional archival appraisal and estimates of enduring value. Massive collection of metadata to generate signals intelligence for terrorism declares that such haystacks of data can speak for themselves. However, giving data a voice requires immense resources, both economic and human, creating more time for particulation. In other words, "increasing the size of the haystack may make it more difficult to find the needle."[43] Data made in networks is relational, despite legal frameworks that delimit and enable secondary applications. The use depends on the stack with which it originates. Importantly, most legal doctrines assume an individual right to privacy when protecting communications data, but infrastructures transmit personal data but do not consider metadata to be personal or private because of the nature of the stack and how data is generated and moved through it.

Today, cloud computing and on-demand partitioned storage architectures drive the rapid expansion of platforms, which have contributed greatly to the decentralization of data storage, distancing it from its original contexts of use. From the 1970s until the turn of the century, in-house computing was on premises in workplaces. But outsourcing in the 1990s and 2000s allowed for virtualized offerings at increasing scales and ease of handoff for wirelessness. As we moved to outsourcing cloud storage, appraising streams of data before it is ingested became less promising. Instead, those organizations could began collecting everything with a "wait and see" approach that the data collected will store answers for future use cases. Instead of collecting data archives to produce knowledge about the past, the semiotics of big data archiving came to be associated with haystack-driven ways of knowing such as predictive intelligence.

The infrastructural ecology of how data circulates through stacks, and then to its final resting places, should not be assumed to be a crystal ball. Nor should the haystacks of big data that can easily accumulate be assumed to have long-term, enduring value. Haystack ways of knowing have limits to the means of retrieving knowledge and making sense of it. Both depend on where data is stored and how it is managed. As others have noted, "storage is a critical moment in the circulation of capital."[44] Where data resides—where it lives once it is collected—is a critical site for examining the flow and friction of valuation and is an occasion for traditional appraisal decisions in terms of costs, value, and evidential use.

Platforms and most organizations are defined by the demand to create data haystacks, and to a certain degree, the imperative to collect data for prediction drives the rapid expansion of platforms altogether. Much like state signals surveillance programs, communications metadata are extracted from user activities and interactions, and the resulting data archives are foundational to the business of platform expansion and network effects of online technologies. As contexts of big data archiving increase and spread, we see a reliance on relationality and interoperability of metadata applications to support both transmission, collection, and reuse. We also witness an increased granularity of data as more and more action that can be represented as micro movements to be further observed and gathered. As such, there is more emphasis on power and control in managing access and turnover—or the handoff points of exchange because of the potential in collection and providing access for reuse and redistribution. In managing the standards and directionality of collection as platforms do, there is more power and value in storing the data after it has been collected than in making sense of it or appraising its value before amassing it in big data haystacks.

A final hurdle for haystack ways of knowing is documenting and maintaining the purposes of its originary context—the provenance information from where data has first been collected and for what purposes. In part, this has roots in the standards with which the data is represented when created (through metadata), but it is also in how the data, once collected and made accessible, is transformed into meaningful evidence for specific purposes or to support particular processes in systems. These circuits of power rely on the original context of data creation for meaning, but the relationality from network handoffs without content misses much—or, ironically, may gather too much so as to be meaningless. This is where we misunderstand that more data necessitates more knowledge. It also explains why the CDR program worked under the law but was largely useless in predicting future events or identifying terrorists. Further, issues of data integrity from the telecommunication companies' handoff to the NSA shows the limits of remanding provenance to metadata but not original content—a robo call can look the same as a call from a friend or a person of interest. Once taken out of the original context, gathered, assembled with other sources, and stored in a big data archive, CDRs are difficult to classify without more contextual knowledge of past action.

The value of data changes over time. This is true for my call data as it is being transmitted through the stack, once it is collected in my screen-time summary report, or when it appears dormant in last year's T-Mobile billing records of US residents. Assembling call data records in massive haystacks assumes that the archive will accrue value in the potential to predict future events and human activities. Here, we see that big data can accrue value like interest, or it may not, depending on how it is used. To be sure, metadata from call data records can indicate whether I'm stressed or that I am lost while traveling overseas. Once collected and accumulated, it can tell us I've been a loyal T-Mobile customer for more than twenty years. Details from my text messages can show business accounts that I have set up dual-factor authentication or reveal my political associations if I'm on a super PAC list of Texas voters. The example of the destroyed CDR database and ending the collection of telephony metadata shows both the similarities between surveillance efforts and platforms that are built on business ontologies of personal data collected from networks and the distinct features of the new data archives made up of metadata. But the turnover of collected CDRs speaks to the power of metadata to assemble a big data archive quickly can reproduce limitations as well.

The CDR program succeeded in extending the dataist logic of haystack ways of knowing by surfacing the limits of context without content. Because big data archives do not ensure evidence and rely on metadata interoperability, they sponge up more resources needed to assign value and assert reliable, authentic meaning. Ultimately, the efficacy of the CDR project was a failure because a lack of enduring value, but it is a valuable illustration of making meaning with big data sources and the risks of epistemic subversion with total storage approaches to archiving it without appraisal—which is why the NSA's announcement from June 2018 to destroy the archive is so unique.

The program ended and the data archives were destroyed not because it was illegal to collect bulk metadata. Instead, it was destroyed because the data could not be used legally in the ways it was assumed to bear evidence to the actions of people coordinating violence via cell phones and networks. Susan Landau and other privacy scholars speculate that quality issues have to do with bad data ingestion and that the telecommunications firms themselves handed over data that was improperly caught up in the bulk dragnet. Readers who have been longtime cell phone owners may also recall that it was around this time that robo spam calls exponentially increased on cell networks in the US.

By 2018, it became clear to the agency and outside observers that collecting hundreds of millions of call records had not thwarted terrorism. Most justifications of the program (and ending it) focused on the nature of metadata created and the legality of it being collected. They did argue against the amount generated, which forced the intelligence community to collect even more metadata. Once compliance under the USA Freedom Act shifted to telecommunication providers' managing the transfer for CDR data, the scale significantly increased, as did the potential for ingesting inaccurate or bad data.

Whether collected by the state, telecommunications corporations, or platforms, today's big data archives rely on a logic that depends on accumulating data from stacks that link and layer telecommunication networks, computing and utilities, corporate actors, data brokers, and institutions. The drive to collect massive haystacks of big data is built upon speculative frames that assume predictive analytics and trying to make future value from data representation and integration with metadata.

These big data–archiving processes involve enclosure and rely on data from stacks. But these networked data sources are not frozen in static repositories. Even as big data archives enclose, they are complicated by participation and particulation that rely on metadata standards like CDRs. These ontologies may drift, become brittle, or have limits as useful sources when taken from one context such as networked communication and applied to others such as signals intelligence.[45] Unlike traditional archives, big data–archiving projects lack meaningful appraisal—those processes of valuation that examine the long-term function and evidential and informational meaning that could be found from such collections. It promises a value that is assumed to become, without testing the limits and realities of enduring value.

6

GATEKEEPING ACCESS TO DATA

The introduction of data-producing mobile devices that connected to internet and mobile networks in the early 2000s ushered in a new era of ubiquitous computing that would shape our experiences of archiving machines today. If software applications on mobile devices such as Palm Pilots and cell phones increased personal data management activities to the synchronization and transmission of digital data, then connecting them to the internet would radically transform the means of accessing data once it could be both transmitted and collected across networks. Like the synchronization practices of early PDAs using wires and specialized memory ports, wireless transmission with mobile networks created new opportunities for data management once haystacks of big data could be easily accumulated with entrenched collection processes. These collection and management practices of born networked data paved the way for an entirely new market for platforms to manage data collections that make individuals less responsible for, and aware of, personal data with the rise of platform-driven data access.[1]

As primary drivers of datafication, platforms have become a dominant form of access to data over the last decade for corporate, government, and research stakeholders.[2] Yet, "access" can have many meanings. It can be a way of approaching a place, a means of extracting resources, or a design commitment that drives actions that users can take in systems. Vexingly, access technologies connected to the internet can attempt to encapsulate all these meanings, leveraging their political and cultural significance while paradoxically imposing more limits and locks, resulting in lock-in that inhibits data portability, preservation, and the production of knowledge.[3] This chapter argues that as collections of personal data moved from management in individual and state contexts to third parties such as platforms and data brokers, personal data became lucrative. Values emerged in both accumulating and managing personal data to access and redistribution that platforms incorporated and monetized as commercially available information. Platforms that generate, manage, and redistribute data now influence trust in information, the means of public access, and scientific knowledge production. This move to platforms managing access points to the data collections of human activity *from* platform interactions would influence the means of accessing data and the rules of engagement. Platforms now dominate access points, gatekeeping the value of such access to institutions, corporations, and the public seeking to use data created from networks.

Attaching "data" to "access" therefore associates a particular historical object of the recent past with the multiple meanings of access imposed by platforms in their services.[4] As such, our ideas about data—especially personal data and how it should be managed for access—is now designed by platforms as prime actors in ways that information institutions, such as government and university data banks, had guided in decades before. I contend that today, data access is now welded to our understanding of platforms and their logics, both incorporating and subsuming prior grammars of action by providing points of ingress and egress.

Platforms use access technologies to create multisided market connections between individual users, advertisers, data brokers, and software developers among other stakeholders. Importantly, data collections from platforms are accessed through "gates" that only they establish. As powerful gatekeepers, platforms have not replaced prior access regimes wholesale but rather have extended and refashioned them. As many others have argued, establishing these new access points is political.[5] In other words, the value of data access manifested in platforms comes from earlier access regimes but perhaps with different motivations, such as profit, lock-in, network effects, and reputation building.

As we saw in chapters 2 and 3, access to data collections is crucial for organizations and communities for engagement, accountability, and governance. And reliable data access is dependent on standardized formats that are both interoperable and mobile. This allows for programmability, networking, and recordkeeping that support communities of practice—people who may be geographically distanced or who share computing resources. It is also incumbent on data stewards (whether creators, managers, or researchers) to publicize these standard qualities openly, to make accessibility points known to intended users. But access to data archives isn't just limited to technical projects and standard formats and publicity. Both chapters 4 and 5 show how real-time data access can be enrolled into new forms of social organization and circuits of power—from signaling workplace professionalization "on the go" syncing with mobile devices to surveillance programs that rely on cell phone records from telecommunications to secure the state.

The management of personal data can be refigured, taking on new significance. This chapter aims to show how all platforms have established techniques of distancing by enacting access to data, which I refer to as "gatekeeping." In their gatekeeping, platforms enact varied distancing techniques with access points, bringing some actors close to data while pushing others further away.

There are three access arrangements to data that illustrate the gatekeeping power that platforms now have over knowledge production. The first is platform partnerships that corporations such as Meta and X (formerly known as Twitter) have publicized to share data with institutions for digital preservation, which typically rely on APIs. The second are public access projects called "data archives" and "libraries," where datasets are gifted to the public by platforms, often in transparency efforts related to political advertising. The third is data integration services that involve metadata and standards development. The power of gatekeeping to platforms can be found in these access points, which rely on asymmetry, obfuscation, and control. Ironically, these platform access points benefit from the positive perception of "public access" that was developed over centuries by libraries and archives in building trustworthy access. Below, I have selected cases that illustrate how these three access points entrench dependencies for users, scientists, journalists, and federal agencies that further the reach of platform-driven data management.

For contemporary digital cultures, data access deeply impacts our means of personal identity development and building community memory, among other economic, political, and social relations. As a social value, access can materialize in information technologies in a variety of ways, engendering different grammars of action that carry over time and across systems. As a value inscribed in information technology, access deeply influences practices of reading, education, and privacy.[6] So, considering access points governed by platforms as they enforce gates—both the means with which data is created and made accessible and the means with which it is redistributed by platforms—helps us approach "access" not just as archiving machines made up of technical processes, services, or software. Rather, examining platforms' gatekeeping tactics shows how managing access to data has political consequences in social, cultural, and technical contexts that shape how the knowledge we produce is trusted, and how we will access it in the future.

A HISTORY OF TRUST

There is a long history of trusting information institutions such as libraries, archives, courts, and state bureaucracies to ensure individuals' access to information.[7] This trust in institutions has been developed over centuries. The turn of the eighteenth century saw a transformation in public access to information, when Europe's existing monastic and parochial libraries increased their capacities for browsing, borrowing, and public use of collections.[8] The members of these early European collections mainly held institutional or private organizational affiliations and were most often wealthy white men.

In America, the public library movement began in the late nineteenth century[9] and, alongside nation-building efforts, was fueled by the belief that access to information and spaces of self-learning would create a democratic public capable of making "informed decisions."[10,11] Furthermore, sites of accumulated knowledge intended for public access, such as libraries and research universities, adopted rights to read, rights to access information, and the right to privacy when accessing information resources. In the US, ensuring privacy is central to the intellectual freedom that libraries encourage when supporting open access to information.[12] Thus, access is core to defining information institutions, a critical mandate to public institutions such as libraries, and central to why they are worthy of trust according to library and information scholars, policymakers, and more.

Nearly 150 years after information institutions began to institute the value of public access to their collections, advances in technology, organizational change, and political landscapes still influence the realities of "access" in contemporary information environments. Institutions such as public libraries now provide access to a host of information objects beyond books and paper-based special collections, from audio books and streaming content to tools, emulated software, and hardware such as 3D printers.[13] Librarians and archivists themselves have begun to explore how traditional collections should be both conceived and theorized as data.[14]

Over the last fifty years, the migration of data storage from paper to tape and then to magnetic media and solid-state storage coincided with the appearance of data access movements.[15] Digital management made data easier to copy, store, and retrieve. More and more, meaningful public access to data is being transformed and subsumed by platform logics and neoliberal governance of corporate data access. Increasingly, the discourse of access as a value is framed through open access and open science mandates that are shaped to platform-driven access too.[16]

The stance that data access benefits the broader public was largely driven in the US by the rise and saturation of social media—at least temporarily. In the first two decades of the 2000s, early social media networking platforms such as Twitter (now X) and Facebook came to be characterized in the United States as a "public square," by the Supreme Court and by the 2010s, laws were passed to ensure that their use was constitutionally protected.[17] During his administration, President Barack Obama promoted a culture of open data in the US federal government and expanded access to publicly funded research.[18] The Obama administration developed initiatives, including an open data policy and the Project Open Data Metadata Standard, and instituted Executive Order 13642, mandating all governmental information be both open and machine readable.[19] In 2014, the United States Digital Service (USDS) was established to oversee digital government services and ensure that they would operate more like platforms.[20]

It is difficult to underestimate the breathlessness with which social media services and platform logics were enrolled into government data efforts throughout the Obama administration. During the end of President Obama's second term, when the White House administration opened their social media data archives for public access, the USDS managed much of the digital transition before he left office.[21] Beyond federally oriented policies promoting public access to social media data and digital government data, the Obama administration encouraged American corporations to form "information sharing and analysis organizations" for the purpose of spurring innovation and technology development.[22] Much of the open access momentum for publicly funded research and government data established during the Obama era coincided with high-tech companies providing access to their platform data through software development kits (SDKs) and application programming interfaces in the mid-2010s.[23] APIs allow software to connect to platforms. Through them, developers can build extensions and technologies "on" the platform, in the architectural sense of the term.

Early permissive platform APIs ushered in an ecosystem of software development that pushed the features, optimization, and customization of social networking features forward, allowing platforms to collect more data—from both users and the developers leveraging API features for development. For example, URL shorteners were introduced to Twitter by developers using the early versions of their API. The shorter links created by TinyURL and bit.ly opened more room in tweets for human-readable text. Third-party dashboards were developed using the Twitter API with different interfaces for users to monitor streams of lists, hashtags, and breaking events. Since 2006, newer developer APIs with more stringent authentication, less access, and more data limitations have been released.

As such, new versions of APIs create dependencies in the ecology of platform development that impact data access regimes today. Many platforms have histories of releasing permissive APIs to draw in potential software developers with access to user data and then roll back permissions and limit data access once developers are locked into the development environment. Such SDKs and APIs affect whole software-driven development economies, such as the Apple iOS and Google Android app stores, in which app developers creating apps for mobile communication experiences are subject to the app stores' policies.

Open access models such as APIs can benefit corporations with various consequences not only for users who provide data but also for third parties that develop software or advertise using the platform. If platform technologies are designed to be market-driven multisided access points to user data, they incentivize the secondary use of data access by promoting innovation through data exchange, all while creating asymmetric access to these resources with social, political, and economic consequences for redistribution and reuse.[24] For example, the Apple iOS and Google Android app stores both govern how mobile app developers can engage with and use the personal data collected from iPhone and Android handsets. Both stores also implement a cost and fee structure for developers to sell app services available through the app stores.

APIs have become the normative access regime for platforms that leverage capitalized data access models for advertising, personalization, and predictive analytics. "Opening" platform data to data brokers selling targeted advertisements and monetizing content based on access to user data is the primary business model for the largest corporations such as Meta, Microsoft, Netflix, and Google.[25] Although other data techniques exist for accessing data from platforms—such as web scraping, purchasing historical and commercially available social media data, or screen recordings—most of these are prohibited by platforms. Instead, their terms of service drive those who want to access user data to platform-hosted APIs.[26]

The promise of opening and sharing data with APIs, combined with new access encounters managed by platforms, shaped the governance of using aggregated sources of personal data and articulated possible means of innovation as well as the economic incentives of platform approaches to managing data for much of the 2010s. As a result, technical affordances of opening data and open access values generated new opportunities and public expectations for access to data collected at the population level. The ease of access that came from platform APIs created a "Data Golden Age" for scientific researchers, journalists, and academics to collect swaths of human trace data that has indelibly shaped both scientific knowledge and research data management.[27]

I provide this brief historical background because platforms now dramatically influence individuals' and society's ability to access information as well as our rights to data privacy. Platforms have adopted new roles as gatekeepers to our digital culture of data as well as governors who administer the rights to technologies used to access it.[28] The rise of platforms' data management and public access approaches changes what we know and expect data access to be, and the positive sheen of open data access from the early decades of the 2000s has become sullied by the realities of weaponized misinformation, disinformation campaigns, and widespread election interference.

THE RISE OF PLATFORM PARTNERSHIPS FOR PUBLIC ACCESS

To examine the power of gatekeeping that platforms play in shaping access, it is helpful to look at the ways that platform companies such as Google, Amazon, and X (formerly known as Twitter) portray their role as data access providers as well as their efforts at public access and data-focused partnerships. Many social media platforms have long histories of providing data access, but some, such as X, have recently started charging for access to increase revenue.[29] Platform data is of public interest because social media networks such as Facebook and X play major roles in our information environments—from government communications to emergency services and public health messaging to breaking news. Most models of platform data access begin with public partnerships and public APIs.

One of the most active platforms in release data publicly has been the company formally known as Twitter, now X.[30] The firm's efforts at public data access reach back to its early days of development when it became clear the platform would be driven by breaking news and serve as a kind of public square. As social media came to be known more as online public spaces, public–private partnerships began emerging, particularly when promoted by institutional and government agencies in the US, Canada, the UK, and the EU. In 2010, in a much-lauded partnership during the Obama administration, Twitter announced that the Library of Congress would accept and preserve all public tweets dating from 2006 to 2010.

The Twitter archive project was considered a data gift to the Library of Congress that would come to stress test the institution's capacity to preserve massive amounts of digital materials and web-based data collections. Scaling the digital preservation project proved to be both organizationally and technically difficult. In 2013, the Library of Congress announced a revised the plan for approaching how they would preserve social media data derived from platforms to other web archiving projects.[31] By 2017, the project was paused, and the library declared that the collection would be embargoed until further notice.[32]

Despite this early failed public partnership, Twitter proved persistent and a corporate leader among social media companies in their continued public data access. The company continued efforts at public data access with an API solely dedicated to research and programs to share disinformation operations with governments and corporations until its 2023 change in ownership and rebranding to X. The platform API model for research access has its roots in engaging software developers to use platform data to develop and extend offerings.[33] Although APIs have not always been intended for scientific research, they do support iterative experimentation with user data and software.

APIS FOR PUBLIC ACCESS

Although other data partnerships such as the Library of Congress Twitter archive exist, the most enduring method of providing access to public data has been via platform APIs. There are many different API models for providing access, and it is necessary to clarify that APIs aren't just for commercial platforms and app developers alone. Many different data-rich institutions, such as libraries and state agencies, have used them for decades to federate metadata records connecting digitized collection resources available for public access.[34] Despite APIs using many types of access encounters, I contend that social media platforms' API models for advertising and for developers have influenced many of the public access data projects that digital researchers rely upon most.

The primary way that social media companies provide access to bulk platform data is through the APIs directed toward developers. Twitter announced its first API to developers in 2006, following the release of earlier platform APIs from eBay and Flickr. Like many early APIs, the first Twitter API was permissive, with broad access capabilities and few authentication protocols. These capabilities allowed developers to gather lots of platform data as well as modify their interaction with the interface. But what data can (and cannot) be gathered with APIs is important to consider when evaluating the impact and purpose of the public initiatives because they should be weighed against platform interests—services such as protecting their users' privacy or engaging with their developer base.

Relying on API access models can create unintended consequences for various stakeholders, including the public, government agencies, and news organizations. Deletion and moderation from platforms also threaten access to those officials and organizations who are responsible for preserving official records and communications data under the law. For example, when President Trump was banned from Twitter for advocating violence during the January 6 riots, the National Archives and Record's Administration was prevented from gathering the president's tweet data.[35] In a letter to the US House Committee on Oversight and Reform, then archivist of the US David Ferriero explained that other versions of Trump's tweets exist and were collected by "nongovernmental sources." However, NARA "[does] not consider them as official Presidential records and ensure the completeness of their captured account data."[36,37] Trump's banned account and the missing tweets illustrates a limit of the API access model. If data is removed from the platform, it is no longer accessible on the public web or through the platform's APIs. Trends in content moderation, changes in political advertising, and platform manipulation are not API accessible if posts and data are removed from the platform. Still, there may be historical, empirical, or legal significance to preserving and providing access to this data.

Critics generally argue that developer APIs are "black boxes" that prevent kinds of use (and users) that may not reflect the information access values that traditional libraries and archives typically espouse (e.g., privacy, provenance, and authenticity). Such asymmetric information access that limits reuse, future sharing, and tertiary redistribution of the data extracted from APIs signal both active distancing and a kind of constant, immobile wedge for scientific inquiry. Data extracted from APIs is typically from specific and brief periods of time. So, there are also limitations on the claims that researchers can make about behavior, generalizability, and replicability of research based on social media. As access points, APIs may also limit what questions researchers can ask, creating an agnotology of platform behaviors such as media manipulation, departure, or nonuse.[38]

In these ways, the platforms' role in policymaking and governance of how the platform is accessed with developers' APIs also impacts data access in downstream contexts of knowledge production. Many academics and investigative journalists have discussed how developer APIs are both a boon and curse to researchers attempting to access platform data to research online behavior and the influence of social media technologies.[39] Few of these research concerns focus on the downstream impacts of data collected with platform APIs, such as future access and digital preservation outside the platform.

AD LIBRARIES AND DISINFORMATION DATA ARCHIVES

Two years after the discovery of widespread election interference in the 2016 United States presidential race and the Brexit referendum vote in the United Kingdom, platforms began to provide public access to social media data and political advertising data through ad libraries and data archive tools. Social media platforms characterized these data access points as public goods that would increase transparency into information operations and political advertising. Facebook, Twitter, and Snapchat released these resources between 2018 and 2019 and rolled out more policies regarding political funding disclosures to identify those who purchased advertising. These ad libraries and data archives indeed provided unprecedented public access to social media data. However, they are also prime examples of the gatekeeping tactics that platforms use to create both asymmetry and dependency in providing access in the name of public service.

Much of the platform data intended for researchers and journalists in these transparency programs involve refashioning or implementing new APIs for access. For example, Facebook's Ad Library features a web-based search tool, a downloadable report with structured data, and a dedicated API through which users can run queries to retrieve structured data.[40,41] Each connected resource includes advertising data from Facebook's other platforms, including Instagram and Facebook Messenger. However, political advertising was banned, and Twitter's Election Integrity Archive aimed to document political speech and state-sponsored manipulation campaigns. Even as manipulation campaigns targeting elections continue to flourish, the platform periodically publishes additions to the archive, updating known information operations.[42,43] Snapchat's Political Ad Library, which was released around the same time, is a smaller effort that also includes targeting metadata from political ads.[44]

Although API access to advertising data can give researchers evidence about engagement with posts on the platform, this information often does not include targeting metadata, ad spend, or which user account purchased or promoted the post.[45] For this reason, these platform ad archives don't often provide valuable metadata to developers or researchers. In these ways, such efforts are part of a broader strategy: platforms gatekeeping access on their own terms. This is particularly risky for scientific research because there is no way to confirm that the data released from platform access technologies is complete and accurate. Most APIs and terms of data partnerships restrict researchers from sharing their data to reproduce or replicate their findings.[46]

In response to the flurry of ad libraries and their APIs, Mozilla, the nonprofit foundation that supports free software and open standards, published a letter directed at technology companies and policymakers suggesting five API guidelines for effective digital ad archives.[47] Characterizing them as "entrenched," the Mozilla letter argued that the problem of API access models is that they prioritize a closed loop of advertising, diminishing its impact for effective research. One guideline focused on ensuring the availability of up-to-date historical data, and another mandated that a platform API and its data ought to be both "accessible" and "shareable" with the public.[48] Data from APIs and publicly available datasets may be the same data, but like developers' API limits, researcher access comes under the platform's terms of service with limits on future use to the data accessed. Such limitations on use, engagement, and dissemination contrast greatly with the traditional policies for public access upon which libraries and archives were built but illustrate the power in managing collections and access to them thusly. In other words, these efforts are just stunts and serve transparency efforts in name only. Those that control collections control the means of their use.

Ad library tools are indicative of the access regimes that platforms administer in broking user data for targeted advertising.[49] As artifacts of a platform's public interest, they reveal how much more data would be necessary to achieve transparency. Those journalists and researchers who have tried to use the data have found it riddled with errors and lacking documentation. Despite their availability to the public, careful study of ad libraries and data archives show they are designed for limited use cases and inhibit methodologies for deeper analysis. The possible research questions that can be asked of, and the possible answers from, these publicly accessible datasets are limited, demonstrating the difference between public access from information institutions that have enshrined access as a right and public access from platforms that are subject to forces of late capitalism.

Platforms' transparency efforts in releasing advertising information are further limited by ongoing decisions about policy enforcement and shifting regulations in targeted political advertising. Platforms continue to confront emerging data protection legislation as well as ongoing election interference from state-linked accounts. Countries have pursued formal legislation that has shared, aggregated, and disseminated social media data with varying levels of success.[50]

PUBLISHED INFORMATION OPERATIONS DATASETS

In a similar vein to releasing political advertising data, platforms such as Google, Meta, and X have also released data regarding foreign influence and disinformation campaigns called "information operations." These data have been taken down, removed, and flagged as suspicious or manipulative behavior. Seeking to document the role of foreign influence in these disinformation campaigns, Twitter published datasets revealing the influence from Russia's Internet Research Agency and Iran. They appeared as a branch of Twitter's Elections Integrity Hub, which then included a state-backed information operations data archive.[51,52] The initial announcement of the disinformation data archive in October 2018 included the goal of "encouraging open research and investigation" by civil society researchers, journalists, and academics concerned with elections integrity and accountability.[53] Empowering academic insights and increasing public understanding of platform manipulation is consistently repeated in company blog posts announcing releases that were part of broader transparency and security efforts.[54]

Although platform companies aim to raise awareness about "information operations" by disclosing them via datasets, they may, in turn, limit scholarly efforts at representativeness, accuracy, and documentation of these data sources. As a result, many limitations exist. The data might suffer from a lack of context. They might include only active accounts or exceptional public figures. These data shortcomings distort our understanding of the activities of state actors on the platform, which are further mired by the high rates of deletion, the decay of accounts, and the ephemerality and disappearance of posts.[55] Critical scholars have noted that independent outsiders can never confidently verify and authenticate the comprehensiveness of social media data released by firms.[56] Most firms benefit from the initial reputation boost of data philanthropy. So, it functions as a fig leaf for their true intentions. Ad libraries and disinformation data archives generate positive publicity when platforms are under scrutiny but eventually puts empirical research and scholarly reputations at risk. For these reasons, computational social scientists, misinformation scholars, and social media researchers have warned academics from relying exclusively on social media data from platforms alone.[57]

The processes for including data in the ad libraries and data archives are typically outlined in frequently asked questions (FAQs), but there is little documentation about the curation and selection reasoning. Scholars are left in the dark about the number of ads that are not included in, or curated out of, collections, as well as those that have been moderated, banned, and removed. Data journalists and academic researchers have found data from the ad archives to be rife with errors and have cautioned users from relying solely on these resources for platform accountability research.[58] Furthermore, platforms fail to detail how these information operations are identified, evaluated, and de-platformed. Data issues of quality assurance, authenticity, provenance, or even deaccessioning will remain unknown without extensive, internal documentation capturing how these data are assembled for public access.

DESIGNATED ACCESS FOR ACADEMIC RESEARCH

The platform access models of disclosing published datasets and public APIs can be compared to formal academic data programs, which allow for designated access to researchers. Platforms have come up with different approaches for responding to public concerns regarding data access for research. For example, Yelp has released curated platform datasets such as the Yelp Open Dataset, which are intended for educational purposes.[59] Many platforms have created formal programs for academics and graduate students to apply for special research grants and competitive internal competitions for data access, such as the WhatsApp Research Awards or the Snap Research Fellowships.[60] But following social media platform disclosures on political influence and media manipulation—especially after the fallout from Facebook's Cambridge Analytica scandal from improper reuse of user data by an academic researcher—many social media platforms have developed their own set of proposed reforms concerning algorithms, advertising, and engineered content in an effort to self-govern and protect them from manipulation.[61] Platforms publicize these efforts as promoting transparency and civil society oversight, but they conspicuously appear on the back of scandals and public backlashes. This has been especially true for Meta's Facebook, which has a history of announcing data access for broader transparency shortly before executives have been called to testify in special hearings before Congress.

One of the most well-known platform partnerships with academic researchers where Facebook has created hurdles researchers and rolled back data access is Social Science One. In 2018, it was characterized as a "new model" for industry and academics to share social science data. The organization was initially proposed by political scientists, Gary King and Nathanial Persily, in partnership with Facebook, the Social Science Research Council, and Harvard's Institute for Quantitative Social Science to create data access between academia and industry.[62] Working with Facebook to ensure privacy, the goal of Social Science One was to provide researchers' access to company information, particularly elections-related data. Social Science One would act as a kind of trusted third party to award funding and data access to elections data to researchers.

For three years, data access was promised but stymied. In 2020, Facebook finally released the dataset of URLs to researchers, but this was not the original data that was once promised.[63] Such partial data leads to unreliable evidence, creating what sociologists of knowledge call the "agnotology of social media platforms."[64] By circulating massive datasets with missing or flawed evidence, Facebook deliberately creates space for ignorance to grow.[65] Whatever their motivation, there appears to be no meaningful incentive, policy, or market logic for Facebook to accurately share their internal data with the public. Partnerships such as Social Science One that involve both Facebook oversight to data access, uses, and results reveal tactics of control and conflict with research integrity.

Designated platform access programs such as Social Science One that grant teams of researchers special access appear a boon to researchers and journalists with limited computing resources.[66] But for years, Social Science One was subject to Facebook delay tactics. The Co-Chairs of the European Advisory committee issued a statement acknowledging the company's slowdown, saying, "The current situation is untenable. Heated public and political discussions are waged over the role and responsibilities of platforms in today's societies, and yet researchers cannot make fully informed contributions to these discussions. Scholars are mostly left in the dark, lacking appropriate data to assess potential risks and benefits. This is not an acceptable situation for scientific knowledge. It is not an acceptable situation for our societies."[67] Members from Social Science One's European Advisory Committee and researchers who benefited from designated access programs have written about the experience, arguing that delays not only impact selected researchers but also harm the possibilities for broader equitable access to the research community.[68]

Data access for external research has been impeded by Facebook in ways that are well known to investigative journalists and social media researchers. In September 2021, the House Science Committee convened an investigation into researcher access to social media data in order to counter disinformation and media manipulation.[69] Two weeks before the hearing, reporting from *The New York Times* revealed that datasets given to researchers from a Facebook research partnership were incomplete and inaccurate.[70] It wasn't the first time that data has disappeared from Facebook's data transparency and access tools, thwarting researchers and journalists that rely on the platform's data for investigating current events and the spread of disinformation and misinformation online. Shortly before the oversight committee convened, Politico broke the news that thousands of posts from the January 6 insurrection went missing from CrowdTangle, a Facebook-owned analytics tool that allows users to monitor trends across platforms. In 2024, Meta announced plans to deprecate CrowdTangle and replace it with a new tool called Meta Content Library.[71] This transition exemplifies a broader pattern of social media platforms restructuring their public data access programs, notably constraining established research methodologies and analytical capabilities that scholars had previously relied upon for studying the impact of platforms on society.

For more than a decade, Facebook (now Meta) has created hurdles for researchers, journalists, archivists, and civil society groups attempting to access data via more traditional web-scraping methods that circumvent the company's sanctioned APIs. Facebook has blocked journalists' and researchers' investigative tools that track ad targeting by disabling the functionality of browser plugins, disallowing the sharing and datasets among researchers, and cutting off access for controversial research projects. Facebook has even pursued legal action to dispute researchers' claims and the validity of their tools.[72] Time and again, Facebook has shown their transparency and access efforts to support independent research to be perfunctory at best. At worst, they harm scientific and reporting communities concerned with studying the impacts of social media on society. Still, these strategies define the platform's approach to public data access.

IMPACTS ON TRANSPARENCY AND RESEARCH

Each of these platform efforts are framed as data access for the public often in support of transparency, accountability, along with calls for more regulation and guidance from governments. Platforms have a vested interest in supporting researchers examining the impact of their products and algorithmic technologies because they are sites of public discourse and human activity and evidence of societal change. But although researchers value Mertonian norms of science such as communalism and replicability that rely on accessible, open data that can be shared, platforms do not protect the same values of access. Instead, their profit and lock-in goals make them rely on gatekeeping tools that provide asymmetric access to data that inhibits reliable research. Such business strategies are at odds with scientific knowledge and research scholarship. However, over the past decade, more firms have initiated access programs to their data archives directed toward researchers, investigative journalists, lawmakers, and civil society organizations by providing data for transparency into platform operations. Often, these data initiatives follow on from public or political pressure for more independent research to study the impacts of these technologies on society.

Much has changed regarding API access since early platform releases. For their part, platforms now have dedicated research APIs and terms of service that aim to address these reliability and authentication concerns, but as we have seen, they have many limits for empirical research claims. As Tromble warned, "platforms and their APIs have always been proprietary black boxes, never intended for scholarly use. Even when researchers could mine data seemingly endlessly, we rarely knew what type or quality of data were at hand."[73] Boosters to these efforts have argued that more data access for transparency reasons promotes better platform governance and accountability. Whistleblowers such as Frances Haugen believe that the only way to fix Facebook is through regulation that ensures researchers and regulators gain truly transparent access to platform data. But until laws and policies exist to enforce platform data transparency and data governance, researchers will continue to rely on APIs, gifted datasets, and designated researcher access.

Public-facing tools such as Twitter API for research or political ad libraries from Facebook and Snapchat are emblematic of platforms' efforts to provide access to information and data. In their own language, they are thought to increase "transparency" for accountability by gifting data to researchers and providing data archives to the public.[74] But despite these commitments to transparency, the configuration of ad libraries and data archives inhibit access and limit possible lines of inquiry into the platform effects that they are intended to mitigate. Acting as another kind of haystack way of knowing, these slices of selective data for public access can mask the depth of the realities they seek to shed light on. Whether accessing data from public datasets or APIs, researchers and developers alike can only access what platforms want them to see.[75]

DATA INTEGRATION AND THE ENCLOSURE OF METADATA DEVELOPMENT

A third way that platforms act as gatekeepers and create asymmetric access is by distancing data from original contexts through transforming datasets for analysis. Once data sources have been transformed by integration, they are refigured, often in ways that cannot be easily reversed to original contexts or interoperable formats. Data integration services are core platform offerings that serve legacy data to information-rich organizations such as governments and businesses. Data integration can be offered in combination with other cloud architecture hosting, such as AWS, or system administration, such as Workday or Oracle. For organizations with many kinds of records and potential data sources accumulating, the work of integration is not only in extracting data from different sources and formats but also, once gathered, in transmitting it to a storage architecture that stages it for linking and transformation.

This transformation of data integration involves more than just merging two datasets together and cleaning it for analysis. Increasingly data integration services aim to allow for quick retrieval and publication to the web. Leveraging principles of interoperability and integration, it is partly machine driven, but the wrangling of classification and standardization of these resources representing data across systems has traditionally been implemented by human experts, who apply metadata standards for interchange and linking. This representation development and implementation is dependent on accurate data integration and is integral to all networking technologies, such as platforms and communication systems that collect and transmit data. Metadata development has historically happened in public, open, and federated environments governed by standards created by engineering organizations.[76] But for many platforms that contract data integration services, metadata development is an alchemy they wish to protect.

One of the most secretive data analytics firms is Palantir, which offers integration services via custom platforms under contract. The dashboards they produce support police, military units, and humanitarian services of security and management, and they often integrate government information with signals intelligence or video surveillance footage. Palantir's surveillance platforms, like their integration services, rely on data sources originating from siloed collections that are then braided together via metadata applications. Metadata is as crucial a feature of Palantir's activities as it is for platforms such as Uber or Instagram. Data integration is impossible without metadata, which makes data usable through the construction of schemas, graphs, and ontologies to model and represent it.

Massive data collection is often framed as the most potentially harmful process that platforms do. However, recently, data transmission itself has come under scrutiny, particularly in relation to integration and data transfer across national borders and protections for state surveillance.[77,78] According to the firm, Palantir does not collect data. Instead, its clients do. Palantir merely integrates it and makes it interoperable with its platform products such as Gotham, Foundry, and Apollo. Instead of creating APIs that allow for extraction, Palantir's services provide the ability to combine and link different types of data silos using metadata and semantics to link them in providing integration in support of access. So "making metadata" and the enclosure of its development lies at the heart of Palantir's claims that it does not collect data. What it really does is assemble and implement the metadata deployed in data management systems, allowing customers to integrate it with software for data access.

Just as tools such as APIs have created normative access points, contracts for data integration push users toward closed systems development for proprietary data analytics. For data management and software firms such as Palantir, metadata is a structuring force that drives the reach and applications of their core service offerings. But there exist other models of integration that are open and account for change and multiple ontologies. For example, for archives, information about original context distinguishes archival metadata from other kinds of digital collections.[79] For many archival scholars, metadata bears context, which distinguishes it from other digital collections, or data gathered from platforms. There is a profound distinction between traditional archival collections and those of digital platforms. The former preserve context via accurate metadata, whereas with the latter assume linking as a means of enrichment in applying semantic metadata. Platform data integration—linking data semantically and technically—then often flattens context in its aim to represent all actions of the human lifeworld.[80]

Metadata is an essential feature of data technologies, and it shapes the collection, integration, and analysis of any data resource. There are historical, occupational, and popular ways of defining metadata that give us insight into the power and reality of infrastructures used for controlling information now and in the past. For many, the application of metadata in representing data can provide a means of apprehending the datafication of society. For critical data studies scholars, metadata and its development are increasingly a way to make sense of the big data era and distinguish commonalities among communities that use these standards.[81] These moves not only frame how information is represented but also create cascading effects in how data, once integrated, is successively indexed and published online and then reframe what they are meant to be evidence of. According to Roderic Crooks, this is what is known as the "representationalism of data," where "data are distinct from the things they represent, but they stand in mimetic relation to their referents."[82] Andrew Iliadis has argued that these semantic technologies, like standards used for integration, have the potential to flatten concepts in pursuit of universal representation.[83] But when metadata is developed for data integration in closed and privatized environments, it creates immense lock-in for those who wish to rely upon, retrieve, or reuse analysis from data sources. Thus, the representationalism and the semantic standardization of data sources can be elided once data sources are transformed and integrated. When applied to link and represent many sources, metadata then becomes another way in which platforms assert control, establishing meaning of, stripping important context from, and controlling access to data via integration and analytics.

Today, making metadata is how protocols, data technologies, and high-tech firms using networked infrastructures operate. It also guides the operation of communication networks such as the internet and satellite networks.[84] Metadata is essential to networked infrastructures and broadens the reach of the state and digital firms as they connect to data subjects and enforce new data subjectivity.[85] Metadata, and processes for making it, is increasingly crucial to experiences of control. Through metadata, firms can estimate risk, measure communities, and engage in both public and private forecasting of human activity, from voting to migration to shopping.[86] Many library and information studies scholars have examined the power of user-generated metadata from platforms and their impact on future-use cases. They have considered whether user-generated metadata impacts retrieval, reflects cultural biases, or becomes incommensurable with new web-based, linked data contexts. But little research has examined the politics of representation of enclosed metadata applications on platforms.[87] Making metadata over time and across layers of data integration reveals not only the "how" of governing data subjects and new arrangements of surveillance with software but also the role that data integration services play in creating enclosures and opacity.

Data analytics firms such as Palantir build Software as a Service (SaaS) offerings on top of data collections. These offerings rely on metadata and semantic technologies to drive data integration, enrichment, and analytics services to support clients. They promote a unified view of massive data collections, sometimes called "data lakes" stored in warehouses. Like other information firms, metadata ontologies are a core part of analytics business models. However, platform logics of enclosure and lock-in appearing in metadata implementation will likely entrench asymmetric access to data collections even further once they have been integrated into new resources and access contexts. The access that they provide is in creating and assembling data transformation contrasts with API access models that social media firms use, but as platform gatekeeping, it is one side of the same coin.

A contract dispute with Palantir and the New York Police Department (NYPD) from 2017 clearly shows the problem of entrenched asymmetric access with enclosed integration. Firms tend to outsource integration with platforms' integration engines such as Microsoft Azure or Google BigQuery by using APIs and consolidating data from different sources and heterogenous data types. They use data collections to generate and link data, which provides access. Data integration through SaaS often involves decontextualization and format migration as a means of enrichment—a significant shift in traditional information management approaches to access.

From 2011 to 2015, Palantir received many kinds of information that the NYPD collected as part of their service contract for the Gotham platform.[88] From parking tickets to criminal records and car licenses, Palantir integrated these data sources to support NYPD's crime analysis software services. Data-driven approaches to policing have employed software systems for decades, but leveraging platforms for data integration is relatively recent.[89] Some of the sources that the NYPD handed over to Palantir for integration and analysis were public information. Other data, such as criminal records, may be subject to expunction over time. Contract costs for the first few years were estimated at $2.5 million for servers and Gotham licensing. In 2015, when the NYPD planned to discontinue the services and switch vendors to IBM, their Palantir contract ballooned to $3.5 million. NYPD believed that the replacement from IBM would be cheaper and offer users more control than Palantir's dashboard.[90] The status of the analyzed data was disputed by the NYPD before they ended their contract with Palantir as to whether the integrated and transformed data could be accessible once the contract ended. The results of these transformations are made possible by cloud architectures but are often disputed to be another means of locking in customers with customized storage solutions tailored for accessing data once it has been transformed via integration.

When Palantir turned over the analytic data from their integration services, it could not be accessed without using Palantir's software. The extracted data was in a closed format that prohibited the NYPD from accessing it; without an active Palantir license, it was useless. Later, a second dataset was handed off, but the open format contained no analysis or integrative results. Instead, it was simply the original data sources that were supplied to the NYPD over the course of the contracts. The contract dispute between the NYPD and Palantir shows the stakes of data extraction and interoperability once a platform has been used for integration and analytics.[91]

For nearly a century, metadata standards have typically been developed in open contexts where linking for access was the goal of its integration and interoperability across systems, not in private, secret, corporate contexts of firms such as Palantir or Anduril.[92] Data integration services such as the ones Palantir offers are another kind of gatekeeping approach that results in transformed data and asymmetric access to it. They only offer access through integration services that benefit after data comes in and is transformed through cleaning, linking, and integration. Because the analyzed data is not interoperable and cannot be easily extracted from the platform software, format lock-in forces users of subscription services to retain licensing in perpetuity.

Like call detail records made from using cell phones, various types of metadata created by users in public and on platforms indicate the shifting status and the value of data accumulated by platforms once it has been described in a standardized way and integrated for different access purposes. However, unlike traditional archival values of metadata and contextual evidence, platform metadata typically ignore context by flattening it to focus instead on potential future-use applications related to analytics and prediction. Flattening contexts in favor of charting a future for potential reuse has now fixated on the aura and promises of automation. As platform metadata becomes enclosed in private contexts in support of integration and transformation, we must guard against its totalizing force by ensuring open standards for linking data sources.

A PLATFORM IS NOT AN ARCHIVE

The dominance of the platform economy creates new circuits of power that rely on the atomization of data that is collected by networked services, managed for interoperability, and then made accessible through integration. The platformization of web data, as Anne Helmond has shown, makes all sources of digital evidence subject to disruption.[93] Accessing data through platforms that collect and connect data makes it a site of politics as well.[94] Platforms exercise their politics in this space via APIs, data resources such as publicly available datasets (confusingly called "ad libraries" and "disinformation data archives"), and the enclosure of metadata standardization that supports integration and data interoperability. These sites of access are as much social as they are technical, incorporating distancing techniques that significantly shift the means of encountering, witnessing, and making sense of the data we now collectively create.

Platforms constrain access in ways contrary to the historic value of public access found in libraries, archives, museums, and research universities. This results in publics being distanced from the data they create, as the transparency and authenticity of accessed data cannot always be ensured. These new access points usher in a new kind of archival politics reshaping our expectations of access where data can be purchased and management is outsourced, but true public access and ensuring it over time remains a premium. Further, this access regime is an extension of platforms' sovereignty across nations. The largest platforms, many founded in the US, target the newest people coming online worldwide, illustrating how platforms understand their own power.[95] Archiving data and gatekeeping access to it are emblematic of how platforms understand their own functional sovereignty and reach in experiences of the internet, technology development, and historic approaches to standardizing access technologies.

To be sure, motivations for gatekeeping and guarding access to information resources are not unique to early twenty-first-century platforms or social media networks. Information lock-in and asymmetric access have been studied as competitive advantage in business intelligence since the 1980s.[96] Limiting access through data management occurs even further back, in eras of industrialization and analog information paradigms.[97] As the national data bank debate in the 1960s proved, the power in connecting heterogenous sources together is as significant as collecting them for governance and administration. However, gatekeeping with asymmetric access to data collections that platforms manage has established data dependencies for knowledge production that digital researchers continue to debate. Platforms offer data access to third parties because of what it reveals about trends in contemporary digital culture. Educators, scientists, journalists, civil society activists, and policymakers have engaged with these new means of accessing data through gates that generate new dependencies on platforms that shift expectations of not only accessing and archiving data but also trusting its accuracy and completeness.

By gatekeeping data, platforms now dominate its significance in organizational, legal, and political ways that further shape trust in evidence, information, and expertise. First, by organizing new normative pathways of data access, they impact prior paths and funnel users to developers' tools such as APIs. Second, platforms' access points shift the legal and policy implications affecting how data is used and reused by third parties. Many of the data governance issues that platforms face at the national level relate to legitimately providing access to scientists, researchers, journalists, and students. The third constraint is political. In transforming the meanings of public access to data, information institutions, including federal government agencies, have been forced to revisit professional values and first principles such as context and provenance to distinguish their access and preservation mandates from those of platform providers. The state increasingly relies on platform services to support governance such as border security and public health monitoring, but many platforms' access technologies cast them as developer users. By constraining public access to data collections in these dramatic ways, platforms put pressure on existing understandings of data archiving and expectations of access to reliable and trustworthy information.

The access that platforms architect is neither neutral nor a replication of the access that information institutions such as archives and libraries have historically provided. Instead, the data access that platforms produce has reconfigured the politics of cultural memory, as Nanna Bonde Thylstrup has argued, "from the traditional symbolic politics of scarcity, sovereignty, and cultural capital to the late-sovereign infrapolitics of standardization and subversion."[98] Platform access points—APIs, public datasets, and data integration services—carry a new kind of archival politics imbued with platform logics of control, lock-in, and representational enclosure. But these access points are as much social as they are technical, as they shape and remake our expectations of what archival access to data means today.

Platforms have also become venues for communities to document and commemorate trauma, violence, and loss.[99] However, the ways in which platforms convene such publics are notoriously opaque, providing outsiders with little perspective into their construction, development, and evaluation.[100] Despite leaks and whistleblowers, we know little about how platforms rely on data archives that they preserve and manage themselves.[101] This opacity influences how cultural memory becomes instantiated on the platform, mediating long-term accessibility and shaping possibilities for digital preservation.

Since 2006, the very fact of these access tools' existence has been heralded by journalists, policymakers, and especially by the platforms themselves for their potential to improve the accountability of data-driven work with interactions online, not just for social media networks but for innovating gig work, software development, and building automated technologies with models trained on public platform data. But realistically, public–private partnerships and public datasets for platform transparency are in name only—although they are well publicized, they have been found to be incomplete and unreliable. Public access, well-defined scholarly research communities, and government open data initiatives promote information access that the public can trust to be authentic. By contrast, platforms' access points provide access to data from a liminal position that offers only an impoverished sense of "transparency."

The way in which platforms integrate and make data accessible is not well documented or contextualized as traditional public access that information institutions such as archives have been. Indeed, platform-based public data access and data integrations lack oversight and robust measures of accountability for data governance. The irony is that the limits of these access tools, described in this chapter, prove how much *more* transparent platforms offerings could be. Even so, evidence of the reach of harm, lack of policy enforcement, abuse, and manipulation that occur across these platforms continues to be discovered by researchers and journalists. Such instances are evidence of the gatekeeping power of platforms through management and preservation of data for access.

These platform access points also reveal the power of governance, sovereignty, and state censorship campaigns have over access to digital cultural memory. Today, platforms take on the mantle of authority in connecting billions of users to information. They see themselves as providing access to data in ways in which public institutions have historically been responsible. But data from platforms alone is always partial. Policy researchers such as Leerssen and colleagues have suggested that platform ad archives' transparency efforts could be improved. They argue that comprehensive public archives of commercial and political ads would not only benefit governance in advertising policies around the world but also redress information asymmetry found in these publicly available data.[102] Other scholars such as Napoli have argued that such data should be considered a public good, with the potential to help govern platforms.[103]

The access model of social media networks that allowed developers, advertisers, and researchers alike to access and reuse collected data via developer APIs made previous models from information institutions appear slow and stale. But the concessions to platform access models facilitate easier distribution and greater reach and increase the capacity for monitoring and surveillance capture of human activity, impacting privacy. Asymmetric access is not unique to collections of aggregated personal information alone but can be understood as a predictable outcome of information monopolies already present in science, education, and academia, particularly among publishing and public access to information.[104] These current realities might also account for why so many computational social scientists continue to partner and use platforms for data, despite the many well-documented research partnership failures.

The last decade's shifts in how data circulates with platform access reveal how platforms have reshaped popular understandings of data access and assembling archives for research and scientific inquiry. Social Science One was a high water mark for academic–platform partnerships, arriving at an inflection point for social media companies reckoning with misinformation and media manipulation impacting elections and warping public trust. But much like Twitter's archive gift to the Library of Congress the decade before, the program has largely been revised and has stalled due to Facebook's inability to hand over data.

The failure to hand over and provide reliable access has often backfired badly because the siloed data management at scale is a feature of the platforms' internal data engineering approach. Some accounts suggest that providing the data that has been promised is not possible because the company itself is unable to know where all user data is stored. According to leaked internal documents and court testimony, Facebook has massive data lineage issues because a lack of documentation has diminished its abilities to control and explain how data is used among systems with different access privileges.[105] Tracking the provenance of data once it is collected is challenged because the data is spread across systems and stored in separate infrastructures. For firms that contract and provide data integration services as part of their core platform offerings, the dearth of data lineage and plans for documenting provenance for public open-access efforts may be characterized as an intentional corporate strategy of gatekeeping access.

The realities of platforms' gatekeeping data access impact communities and institutions in increasingly intractable ways. The problem with their access tools is clear: platforms provide the public with data access to information, but many of these intended access points are designed for advertising and developers building for-profit technologies. Much of these access points offer only a partial perspective at a high cost, as the data lacks comprehensiveness, accurate representation, and contextual information. And much of the curation and metadata necessary for comprehensive meaning is enclosed and proprietary, whereas the data that is made accessible may change frequently from one user to another because it is guided by shifting internal governance and evolving platform policies. These new platform ad libraries and data archives appear to stand in contrast to earlier data access efforts typified by developers' APIs, web archives, and platform data for purchase.[106] But each involve gatekeeping access where platforms manage data and guard possible uses.

Gatekeepers designate authority through distancing techniques—through the churn of platform service updates, the lock-in of APIs and development tools, and by making metadata in enclosed spaces without documentation. APIs distance users from access to platform data by throttling it. Then, platforms block other forms of access, such as web scraping, and funnel potential access points to APIs. Platforms further shape access by offering corporate research data gifts and special followships, which typically result in incomplete datasets or less access than promised. Platforms ultimately aim for selective data integration by enclosing standards and semantic technologies for specific contexts such as surveillance and advertising but not for open or public access.

This chapter began by arguing that "access" can have many meanings. When applied to gatekeeping, access can designate authority, enable privileges or rights, and even enact the governance by proxy.[107] Platforms can provide certain types of access to data, but they do not offer the comprehensive public access that information institutions provide. Platforms' gatekeeping tactics reveal limitations to accessibility, while platform-provided access tools fall short of traditional transparency and authentic goals of archives to preserve trustworthy evidence. Coincidentally, however, platform gatekeeping tactics do mirror historic modes of archival power, where authority is asserted in controlling access, emphasizing security and surveillance over accountability and transparency.[108] Indeed, the characterization of archives as prisons—where information is controlled, and power is exerted through limiting access—resonates strongly with the platform gatekeeping we observe today. Platforms act as both barriers and gateways to archiving because they commodify, circulate, and provide access to data in ongoing collection contexts. Although platforms cannot be archives in the contemporary sense of access and transparency, the power of platforms over gatekeeping will continue to shape the politics and practices of accessing digital cultural memory.

EPILOGUE: DATA AT REST

When I started this project charting the history of data management, I began with a question: When did archiving become a process that computers can do? This basic question gets strangely difficult to answer because of the many meanings of the word "archive" over the past seventy years. Such polysemy reveals a politics. I came to discover that the denominalization (or verbing) of archive as a technical process to digital data occurred as growing data collections were accumulated and needed to be shared among organizations, firms, and data banks.

In the 1960s and 1970s, before personal computing and the internet, archiving data meant to physically store it away from machines for future access. In its early meaning from data processing with early computers, "archiving data" meant to back up data, take it offline, and put it to rest. Throughout this book, I have argued that the significance of this change can be seen in the externalization of digital data from computers into standardized storage formats. These processes, both technical and social, were incorporated as distancing techniques used by firms, organizations, and individuals to manage larger and larger tranches of data.

During the middle ages of computing in the 1960s, new storage formats such as magnetic tape reels allowed for data handling approaches that subsumed earlier analog and manual processing with punch card data. Magnetic reels and tape drives discussed in chapter 3 not only created more robust storage to manage and move data but also allowed for faster processing, handling, and segmentation. These data management methods, now foundational to today's data-driven analytics, include measurement, collection, transcription, validation, organization, storage, aggregation, retrieval, and protection. New digital data formats, such as magnetic tape, allowed organizations and firms with access to computers and software programs to manage different types of data structures. With machine standardization and standardized data formats, system interoperability facilitated broader data dissemination. And as machines and software became increasingly interactive, data became more programmable and easier to combine, link, and share among different groups of people with access to minicomputers and tape drives to process data.

The formats themselves enabled significant advances in representation because of these indexical qualities. Encoding data sequentially on a ribbon of tape as long as a basketball court obviated the risk of a stack of cards falling over and losing their stacked order, but magnetic tapes were also cheaper and more resilient too. Tapes contained more information and took up less space than punch cards when stored on shelves because they were more compact and lighter in weight. As data could be stored in smaller and more stable formats, the possibilities for copying, sharing, and moving collections of data increased. So, transmitting more data became possible by mailing tapes through the postal service and carrying them by hand. Sharing copies of tapes delivered by hand allowed a community of developer enthusiasts to flourish. These social network practices relied on a durable standard with powerful representational capacities, ensuring data could be stored reliably and represented across time and place.

Once formats for reading, writing, and handling data had been standardized, data could be represented and processed across interoperable contexts—interoperable among different computers and social domains too. Standard formats allowed data, code, and software to be processed by different machines, enabling their spread. Although many of these early computing communities were complementary domains, such as telecommunications labs and university research hubs discussed in chapter 3, they had conflicts over proper access, such as the Unix diff tape that was used to spread updates of version 6 freely.

Standardized storage formats and their interoperability across machines marked the beginning of a new era in data interactivity and representation. As I argue in chapter 4, machine dependencies and indexicality would increase our interactivity machines, also leading to new systems of representation such as electronic metadata, file directories, and interchange. In the process of being formatted and gathered for various access contexts, new capacities for representation with automated metadata allowed for syncing data across multiple devices just as network transmission began to appear. Early networking and time-sharing utilities further expanded what data archiving means and what meaningful "access" has become for different stakeholders as transmission came to be an expectation of digital data.

Until the early 2000s, most computing power was located within the physical walls of firms. Personal computing in the 1980s brought machines into homes and schools and onto our desks. Eventually, we would carry devices close to our bodies, in our pockets, purses, backpacks, and cars. As data flowed through mobile phones and software apps tethered to cellular networks, it became born in networks. The ascendency of third-party access began to drive intensive collection techniques for born networked data. Once data collected are named as archives, given archival potential, or imbued with archival qualities, this configuration is hard to "unsee." The implications for cultural memory are clear. Access tools such as CDR formats and APIs allow third parties to connect to big tech platform data, build software applications, and feed surveillance programs, turning user data into productive resources integral to processes of datafication.

Over the past two decades, access and accumulation contexts turned individual users and their most personal information technologies into sources of an expanding data production process whereby users sync data and leverage networks poised to support the production of a cascading data supply chain. As this monetization model of collecting data and delivering access appeared with internet-connected devices and information services, avenues for becoming subjected to data archiving processes with personal information technologies such as mobile phones and social media apps increased and intensified. The standardization and transmission of user data from mobile devices and wireless networks combined with APIs from platforms turned these firms not only into producers but also aggregators and managers of massive datasets. These data archives were extracted from millions of users, representing their activities into smaller and smaller slices of data.

As more and more AI applications are publicly released and we find they leverage public information published and accumulated online, the full reality of this model of data archiving is laid bare. It appears now that *only* computers can archive our data, and that once collected and subjected to this data archiving model, our data can never be fully taken offline, brought back, or fully known. Such extraction and accumulation rely on not only extreme machine dependencies but also third parties for transmitting and storing our data. Increasingly, we can only derive meaning through metadata and automated processing techniques, leading to more intensive dependencies on archiving machines controlled by the firms that manage and link data. Crucially, many products and services developed by third parties aim to connect and integrate data for future training applications rather than to develop products or services for users to access their data, control and protect it, or put it to rest. This distancing is often echoed and reinforced in the hype around popular AI tools. The more data that we create, the more it is opened and made accessible to be scraped and fed into datasets that train machine learning models for AI applications that drive ChatGPT, CoPilot, or DALL-E.[1]

During the final days of writing this book, I received an email and a text message from T-Mobile informing me that work on their network infrastructure had just been completed near me. The two messages were worded slightly differently, but both had the same gist: without notifying me in advance, maintenance had just been completed on the cell tower I use most. Although there always seems to be some kind of infrastructure work going on in my neighborhood, I had not noticed any differences in my mobile coverage or my data application services. T-Mobile has been my mobile internet service provider for more than twenty years. I have tethered mobile devices to the company's networks across states, regions, and continents, day after day, week after week, and year after year. I always keep my mobile phone close. Except for a few feature phone handsets in the early 2000s and on a few long-distance hikes in the Eastern Sierras without coverage, these devices have nearly always been powered on and connected to mobile and wireless networks, poised to send and receive data.

The maintenance work on my most-used cell tower got me thinking about the archiving machines that we unconsciously rely on and use the most often. For decades, I have used these wireless networks to connect, communicate, and belong. Every machine I use—my writing software, mobile phone, and smart home speakers—relentlessly generates streams of data and metadata. This data is systematically collected and managed by third parties, which turn our personal information, automatically archived and born in networks, into profit. Without a doubt, some of these born networked records have found their way to the NSA's big data archives discussed in chapter 5. The normalization of this datafication, driven by the dual realities of our information society, is striking. The first reality, as computing historian Jon Agar has described, is the "constant touch" of mobile devices tethered to the internet.[2] The second is an assembly of archiving machines that are then stacked to support and collect vast amounts of data from this constant touch to networks. The constant touch to these archiving machines results in the pervasive belief that objective quantification can be captured in data representing our behavior, that it can be constantly captured, and that it should be. Taken together, these realities drive an ideology that legitimizes data access for corporate actors, tech firms, and government agencies. But this belief, this dataism, is premised on trusting the firms that collect, manage, and preserve it for access.

Today's dataism involves a longer history of information technology and negotiating access, bound up in principles of data management and automatic archiving. Data archiving is now predominately used to broker access to data for profit or to leverage algorithms for personalization and advertising. There are fewer and fewer ICTs that do not rely on culling user data because using metadata in exchange for information services is a prevailing norm in our network episteme. We stand at a threshold of a time dominated by AI technologies that depend on accessing data archives, where generative AI evangelists and tech firms continue to advocate for access to publicly available data collected from and published to the internet. The limits of accessing these massive archives call into question their power in making knowledge. Data corporations and data brokers are the new archons—the stewards of our memory and memory's merchants.

The power of archives has always been in their organization and control over memory. Whoever has access to archival material exerts enormous control over who has the power to speak with it, generate evidence, create knowledge, and make memory. The rise of platforms' data management and access approaches change not only what we know and expect data access to be but accessing archives of data too. As chapter 6 argues, the platforms' gatekeeping tactics and positive sheen of open data access from the early decades of the 2000s have become sullied by the realities of weaponized misinformation and election interference combined with hollow plays for public transparency with academic data access programs for research.

To understand our access landscape, we need a theory of networked memory that accounts for the public–private networked archives we are now witnessing, participating in, and constantly creating but cannot fathom and rarely access downstream. Big tech platforms such as Meta, Amazon,

Google, and Microsoft have become the new archons of memory, assembling systems of data collection, accumulation, and preservation management that shape the conditions of asymmetric access and, in some contexts, create haystacks of population-level data.[3] Chapter 6 shows how platforms are gatekeeping access to data for scientists, journalists, and lawmakers. These asymmetric access points to platform data help us understand how the collection and management of data can cut both ways and that more data does not always ensure more meaningful or actionable evidence.

Haystack ways of knowing that promise predictive insights require the constant effort of building and maintaining them. These haystacks are intensely resource consumptive with economic and environmental impact. It is why the concept might harden into the new, expansively applied, polarized, and modular form that it now has. It does not fit a simple binary logic of asymmetric access is good or bad, corrupt or authentic—archiving machines can be an instrument of platform power, but they are not simply another name for big tech. Archiving machines are now built into personal devices as well as governments and institutions. They are used to secure borders, to measure pandemics, to participate in civic society, and to simply live and enjoy creature comforts at home.[4]

For centuries, accessing archives involved managing physical objects and paper documents based on ideas of respecting context, existing relationships, structure, order, and form largely due to analog formats and originality. The power of accessing such traditional archives was constrained by physical location and by the discretion of archivists, magistrates, and government bureaucrats who chose what to preserve or discard. In contrast, digital data, once collected, are inherently mutable and mobile, easily transmitted, redistributed from platforms, and perpetually streaming. Data is now born in networks, accessible in so many ways that even leaks and hacks have become a routine way of characterizing access. Yet, the value of archiving data today is achieved through thick regimes of managed and controlled access, as discussed in chapters 5 and 6. The firms and data brokers that dominate data archives grant not only access and reuse grounded in mutability but also the promise of future growth and predictive analytics, separating original contexts from whence data began. By severing data from their inception, the gathering and stitching together born networked data in cloud storage architectures becomes a defining element of archival power in our present regimes of access.

To understand the impact of asymmetric access on conceptions of archiving today, it helps to hold in mind how these conditions of collection and storage were forged together as minicomputers, digital storage, and time-sharing systems were adopted by information institutions, research labs, and universities. To chart what the last decade of asymmetric data access to user data from platforms has meant for researchers and institutions, I turned to an earlier generation of information professionals and preservation technicians that advocated for national standards of data archives at the same time that data collection and storage was migrating to magnetic tape and digital media. Early data bank managers such as Ralph Bisco featured in chapter 2 predicted tectonic shifts in terms of providing access to data once it could be centralized, connected, and digitally formatted. But after the National Data Center proposals failed, it became clear that data access would be negotiated in corporate contexts.

With automated processing and time-sharing with minicomputers in the 1970s, the rise of the database and data interchange catalyzed networking technologies and personal computing devices that became the foundation of the wireless internet that we make today. At each of these junctures, there were people concerned (both boosters and alarmists) about who would own data once it was accumulated, how it could be leveraged over time, and how such evidence could be both made accessible and controlled with data management administered by third-party firms. However, much of these data management techniques aimed to diminish active personal information management from individuals to machine-driven, automated processes. The data privacy fears of alarmists turned out to be correct, but the control over data has shifted to corporations that have extensive reach in collecting and assembling data archives and act as gatekeepers to access.

Contemporary data archiving in corporate contexts is a continuation of historical processes of data dispossession, distancing creators from their data and implementing automated processes of managing it without their awareness. By examining epistemological shifts in people's understandings of information and users' awareness and data archiving through evidence found in late twentieth-century computing technologies, we can develop a theoretical understanding of what data management and access means to the realities of datafication now and in the near future. As we cede the ability to manage our data in exchange for using ICTs, the implications of this shift become increasingly obscured, leaving corporations with unprecedented power over personal data and information we create in networks.

As other historians and critical archival scholars have argued, it is hard to locate beginnings and endings of archival concepts because the discursive power of archives reach beyond their limits.[5] This reach explains why the concept of archiving data might harden into the new, expansively applied, polarizing, and the popular, meaning it has come to mean everything and nothing. So, "archiving" becomes shorthand for cloud storage, for de facto deletion, or that which is sacrificed in exchange for so-called free but essential services like email. In short order, this feedback loop has become entrenched, where archiving is not led by careful appraisal and deliberate preservation of valuable information but rather the reverse: data archiving from products and services leads to the production of more data for further extraction accumulation for machines to access. A user's activity in information systems, and grammars of action such as writing, searching, and posting, enables platforms to collect and gather data to aggregate it and now pull data into AI services.

Platforms understand that the power of managing data is both in its accumulation and storage to control access. Yet, there appears to be little consideration of what should be preserved or for how long. Those firms with the largest data archives leverage third-party access to their profit and benefit, while those people creating data are more likely to become sources of data to these archiving machines. But as I have shown, this reality goes farther back than the first decades of the new millennium. Whereas many scholars have focused on the history and development of privacy rights associated with the government collection and corporate collection of data about people, few have examined the stakes of what this model of accumulation of data from human action means for the ways in which we access evidence, digital culture, and memory now as we see the datafication of all possible human activities processed through platforms. Importantly, these strategies are not limited to the US. Platforms are global in their infrastructural aspirations, driven by the rapacious pursuit for more sources of data streams to turn human activities into assets.[6] As these firms turn our activities into data assets, they accrete power in society by formatting and representing data and in governing access to archives of digital culture as well.

But unbridled collection does not ensure more knowledge, as the failure of big data archives of the NSA's CDR program remind us.

Throughout this book, I have argued that concepts of archives and archiving have evolved over time, not just shaping people's understanding of the workings of bureaucracy and the state but also documenting and preserving the culture of everyday life. Automatic saving to cloud storage architecture today requires a complex regime of access and material resources that prioritize the collection infrastructure of corporations through time-sharing utilities. It also assumes a series of machine dependencies, with indexicality to represent, preserve, and render data accessible.

The relation of data to society, culture, and politics must be appraised with respect to its own problematics—the formats that carry data through time, the metadata standards that are used to make it machine readable and exchange information across systems, and the access regimes used to witness and make knowledge with data. Here, I have argued that with grammars of action and access points archiving machines transform evidence while conditioning the possibilities of knowing various cultures of data.

The meaning of archive is now shifting again with the perdurance of digital data born in networks and managing access to it. But it is less clear how the jussive power of archives made of data can be asserted when data is rarely taken offline or put to rest. Today, the means of archiving data are used politically and commercially to disguise what this existence makes possible and what is being driven to extinction.[7] Here, we are witnessing a new discursive formation of archive—one that is deeply resource intensive due to the extensive infrastructural demands and energy required to operate them. Managing computational processing power and data centers, in the service of amassing archival power, involves gatekeeping access that shapes the boundaries of knowledge and what can be known. The question now is: What conditions exist to break out of data archiving processes, which seem to constantly reform access regimes in an age of platform capitalism? How might we disrupt or dismantle data archiving processes that perpetually reinforce platform power? One way is to map how data is made both accessible and inaccessible, as I have done in unpacking the distancing techniques that have built up around digital data collections since the 1960s.

This book questions how archiving became a verb by examining the computing processes that contributed to the datafication of our society today.

Ultimately, understanding the history of data archiving shows how new power formations generate inequalities by controlling access to digital cultural memory. These distances, once established, cannot be easily recovered. As archiving machines, these computing technologies localized and emplaced data before it could flow through contemporary, networked information infrastructures. If I am right about the future of accessing archives, then it is time to develop alternative ways to connect to networks so that we might reconfigure control and find ways to put our data to rest.

NOTES

CHAPTER 1

1. For more on the power of interpolating digital data, please see Ronald E. Day's work, particularly Ronald E. Day, *The Modern Invention of Information* (Carbondale: South Illinois University Press, 2008); Ronald E. Day, *Indexing It All: The Subject in the Age of Documentation, Information, and Data*, 1st ed. (Cambridge, MA: MIT Press, 2014).

2. Wendy Hui Kyong Chun, *Updating to Remain the Same: Habitual New Media* (Cambridge, MA: MIT Press, 2016); Tung-Hui Hu, *A Prehistory of the Cloud* (Cambridge, MA: MIT Press, 2015).

3. Martin Campbell-Kelly et al., *Computer: A History of the Information Machine*, 3rd ed. (New York: Routledge, 2019), https://doi.org/10.4324/9780429495373; Paul E. Ceruzzi, *A History of Modern Computing*, ed. William Aspray, 2nd ed. (Cambridge, MA: MIT Press, 2003).

4. For more on early network cultures, see Kevin Driscoll, *The Modem World: A Prehistory of Social Media*, 1st ed. (New Haven, CT: Yale University Press, 2022); Cait McKinney, *Information Activism: A Queer History of Lesbian Media Technologies*, annotated ed. (Durham, NC: Duke University Press Books, 2020); Steven Levy, *Hackers: Heroes of the Computer Revolution*, 1st ed. (Sebastopol, CA: O'Reilly Media, 2010).

5. For more on the materialities of digital information, see Paul Dourish, *The Stuff of Bits: An Essay on the Materialities of Information* (Cambridge, MA: MIT Press, 2017).

6. Jathan Sadowski, *Too Smart: How Digital Capitalism Is Extracting Data, Controlling Our Lives, and Taking Over the World* (Cambridge, MA: MIT Press, 2020).

7. José van Dijck, "Datafication, Dataism and Dataveillance: Big Data between Scientific Paradigm and Ideology," *Surveillance & Society* 12, no. 2 (May 9, 2014): 197, https://doi.org/10.24908/ss.v12i2.4776.

8. Piotr Steinkeller, "Archival Practices at Babylonia in the Third Millennium," in *Ancient Archives and Archival Traditions: Concepts of Record-Keeping*, ed. Maria Brosius, Oxford Studies in Ancient Documents (Oxford: Oxford University Press, 2003).

9. John Ayto, *Dictionary of Word Origins* (New York: Arcade, 2011), 34.

10. C. T. Onions et al., *The Oxford Dictionary of English Etymology* (Oxford: Clarendon Press, 1979), 48.

11. Walter W. Skeat, *An Etymological Dictionary of the English Language*, new ed., rev. enl. (Oxford: Clarendon Press, 1924), 29.

12. Robert K. Barnhart, *Chambers Dictionary of Etymology* (Edinburgh, UK: Chambers, 1999), 49.

13. Barnhart, *Chambers Dictionary of Etymology*, 49.

14. Severo Ornstein, *Computing in the Middle Ages: A View from the Trenches 1955–1983* (Bloomington, IN: First Books, 2002).

15. Richard Pearce-Moses, "SAA: Glossary of Archival Terminology," A Glossary of Archival and Records Terminology, accessed January 19, 2012, http://www.archivists.org/glossary/term_details.asp?DefinitionKey=54.

16. "SAA Dictionary: Archive," 2023, https://dictionary.archivists.org/entry/archive.html#ucd3e2d74648b2f74.3d32d4e.15a432086a6.-285a.

17. A. S. Wight, "The EMAS Archiving Program," *The Computer Journal* 18, no. 2 (January 1, 1975): 131–134, https://doi.org/10.1093/comjnl/18.2.131.

18. Ronald Plavchan, "The International Scene: News and Abstracts," *The American Archivist* 42, no. 2 (April 1, 1979): 230, https://doi.org/10.17723/aarc.42.2.r367725p283xt569.

19. "SAA Dictionary: Archiving," accessed September 24, 2023, https://dictionary.archivists.org/entry/archiving.html.

20. Hannah Alpert-Abrams et al., "Post-Custodial Archiving for the Collective Good: Examining Neoliberalism in US–Latin American Archival Partnerships," *Journal of Critical Library and Information Studies* 2, no. 1 (March 3, 2019), https://doi.org/10.24242/jclis.v2i1.87; Mél Hogan and Sarah T. Roberts, "Archiving for Extinction," *Media-N* 19, no. 1 (February 3, 2023): 7–26, https://doi.org/10.21900/j.median.v19i1.936.

21. Wolfgang Ernst, *Stirrings in the Archives: Order from Disorder*, trans. Adam Siegel (Lanham, MD: Rowman & Littlefield, 2015), 29.

22. For more on the relationship between storage and memory, see Wendy Hui Kyong Chun, *Programmed Visions: Software and Memory* (Cambridge, MA: MIT Press, 2011); Wendy Hui Kyong Chun, "The Enduring Ephemeral, or the Future Is a Memory," *Critical Inquiry* 35, no. 1 (2008): 148–171, https://doi.org/10.1086/595632.

23. Philip E. Agre, "Changing Places: Contexts of Awareness in Computing," *Human–Computer Interaction* 16 (2001): 177–192.

24. Christine L. Borgman, *Scholarship in the Digital Age: Information, Infrastructure, and the Internet* (Cambridge, MA: MIT Press, 2007); Kean Birch, "Technoscience

Rent: Toward a Theory of Rentiership for Technoscientific Capitalism," *Science, Technology, & Human Values* 45, no. 1 (January 1, 2020): 3–33, https://doi.org/10.1177/0162243919829567; Rob Kitchin, *The Data Revolution: Big Data, Open Data, Data Infrastructures and Their Consequences*, 1st ed. (Los Angeles, CA: Sage, 2014); Theodore M. Porter, *The Rise of Statistical Thinking, 1820–1900*, new ed. (Princeton, NJ: Princeton University Press, 2020); Alain Desrosières, *The Politics of Large Numbers: A History of Statistical Reasoning*, trans. Camille Naish (Cambridge, MA: Harvard University Press, 2010); Nick Srnicek, *Platform Capitalism*, 1st ed. (Cambridge: Polity, 2016); Shoshana Zuboff, *The Age of Surveillance Capitalism: The Fight for a Human Future at the New Frontier of Power* (New York: PublicAffairs, 2019).

25. Microsoft's Azure, Amazon Web Services, and Google Cloud Platform are among the current leaders.

26. Frank Pasquale, "From Territorial to Functional Sovereignty: The Case of Amazon," *Law and Political Economy* 6 (2017), https://lpeproject.org/blog/from-territorial-to-functional-sovereignty-the-case-of-amazon/.

27. Katie Shilton, "Values Levers: Building Ethics into Design," *Science, Technology, & Human Values* 38, no. 3 (April 23, 2012): 374–397, https://doi.org/10.1177/0162243912436985.

28. For more on the coloniality of data centers, see Yung Au, "Data Centres on the Moon and Other Tales: A Volumetric and Elemental Analysis of the Coloniality of Digital Infrastructures," *Territory, Politics, Governance* 12, no. 1 (2024): 12–30, https://doi.org/10.1080/21622671.2022.2153160; Sebastián Lehuedé, "Territories of Data: Ontological Divergences in the Growth of Data Infrastructure," *Tapuya: Latin American Science, Technology and Society* 5, no. 1 (2022), https://doi.org/10.1080/25729861.2022.2035936.

29. For critiques and perspectives on everything being "an archive," see M. L. Caswell, "'The Archive' Is Not an Archives: On Acknowledging the Intellectual Contributions of Archival Studies," June 27, 2021, https://escholarship.org/uc/item/7bn4v1fk; Gabriella Giannachi, *Archive Everything: Mapping the Everyday*, 1st ed. (Cambridge, MA: MIT Press, 2016).

30. Joy Lisi Rankin, *A People's History of Computing in the United States*, illustrated ed. (Cambridge, MA: Harvard University Press, 2018); Campbell-Kelly et al., *Computer*; Andrea Butter and David Pogue, *Piloting Palm: The Inside Story of Palm, Handspring, and the Birth of the Billion-Dollar Handheld Industry*, 1st ed. (New York: Wiley, 2002); Levy, *Hackers*; Jon Agar, *Constant Touch: A Global History of the Mobile Phone* (London: Icon Books, 2013).

31. The seizure of President Trump's Twitter account data from X will inevitably show the age of when my research tapered. However, I am indebted to several important histories of computing and US surveillance policies, including Sarah E. Igo, *The Known Citizen: A History of Privacy in Modern America*, illustrated

ed. (Cambridge, MA: Harvard University Press, 2018); Jill Lepore, *If Then: How the Simulmatics Corporation Invented the Future* (New York: Liveright, 2020); Yasha Levine, *Surveillance Valley: The Secret Military History of the Internet*, illustrated ed. (New York: PublicAffairs, 2018).

32. Payal Arora, *The Next Billion Users: Digital Life Beyond the West* (Cambridge, MA: Harvard University Press, 2019).

33. There are many other leading archival scholars theorizing contemporary cultures of web archiving and government digital preservation, including Anat Ben-David, Adam Kriesberg, Emily Maemura, Muira McCammon, and Ed Summers.

34. Anne J. Gilliland, *Conceptualizing 21st-Century Archives* (Chicago, IL: Society of American Archivists, 2017); Terry Cook, "'We Are What We Keep; We Keep What We Are': Archival Appraisal Past, Present and Future," *Journal of the Society of Archivists* 32, no. 2 (October 2011): 173–189, https://doi.org/10.1080/00379816.2011.619688; John Ridener, *From Polders to Postmodernism: A Concise History of Archival Theory* (Duluth, MN: Litwin Books, 2009).

35. Karin Knorr Cetina, *Epistemic Cultures: How the Sciences Make Knowledge* (Cambridge, MA: Harvard University Press, 1999); Gilliland, *Conceptualizing 21st-Century Archives*.

36. Sandra Braman, *Change of State: Information, Policy, and Power* (Cambridge, MA: MIT Press, 2009); Borgman, *Scholarship in the Digital Age*; Paul N. Edwards, *The Closed World: Computers and the Politics of Discourse in Cold War America*, reprint ed. (Cambridge, MA: MIT Press, 1997).

37. For more on the social construction of records and other archival theories of value, please see early work from Ciaran Trace: Ciaran B. Trace, "What Is Recorded Is Never Simply 'What Happened': Record Keeping in Modern Organizational Culture," *Archival Science* 2, no. 1 (March 1, 2002): 137–159, https://doi.org/10.1007/BF02435634; Ciaran B. Trace, "On or Off the Record? Notions of Value in the Archive," in *Currents of Archival Thinking*, ed. Terry Eastwood and Heather MacNeil (Santa Barbara, CA: Libraries Unlimited, 2010), 47–68, http://archive.org/details/currentsofarchiv0000unse_d0m0.

CHAPTER 2

1. Digital material still takes up form, space, and energy. For more accounts of how digital data "matters," taking up energy, form, and increasingly environmental resources, please see Mél Hogan, "Facebook Data Storage Centers as the Archive's Underbelly," *Television & New Media* 16, no. 1 (January 1, 2015): 3–18, https://doi.org/10.1177/1527476413509415.

2. For more on early material records bearing physical inscription, see Denise Schmandt-Besserat, *How Writing Came About*, abridged ed. (Austin: University of Texas Press, 1997); Michael E. Hobart and Zachary S. Schiffman, *Information Ages:*

Literacy, Numeracy, and the Computer Revolution (Baltimore: Johns Hopkins University Press, 2000).

3. Jon Agar, *The Government Machine: A Revolutionary History of the Computer* (Cambridge, MA: MIT Press, 2003).

4. It is worth noting that punch cards were not the first stacks of cards that influenced the ability to create paper-based database systems. During the French revolution, playing cards were used to catalog books using the front and back of cards, thus creating two pieces of information in a table. For more, see Judith Hopkins, "The 1791 French Cataloging Code and the Origins of the Card Catalog," *Libraries & Culture* 27, no. 4 (1992): 378–404.

5. To support this chapter's argument, I mostly draw upon primary sources, federal reports, congressional hearings, era popular computing perspectives, and earlier research from data managers on social science data archives and what was called social information processing from the 1950s to the 1970s.

6. In retelling this account, I am indebted to other comprehensive scholarship on the topic, including historians and information studies scholars. Other accounts of the National Data Center proposals focus on the public awareness of government recordkeeping, the efforts of the Census Bureau in providing access to statistical data, and its role in the 1974 Privacy Act. See Lepore, *If Then*; Rebecca Kraus, "Statistical Déjà Vu: The National Data Center Proposal of 1965 and Its Descendants," *Journal of Privacy and Confidentiality* 5, no. 1 (August 1, 2013), https://doi.org/10.29012/jpc.v5i1.624; Dan Bouk, "The National Data Center and the Rise of the Data Double," *Historical Studies in the Natural Sciences* 48, no. 5 (November 1, 2018): 627–636, https://doi.org/10.1525/hsns.2018.48.5.627; Christopher Loughnane and William Aspray, "Rethinking the Call for a US National Data Center in the 1960s: Privacy, Social Science Research, and Data Fragmentation Viewed from the Perspective of Contemporary Archival Theory," *Information & Culture* 53, no. 2 (May 1, 2018): 203–242, https://doi.org/10.7560/IC53204.

7. For more on how digital data storage becomes invisible as it moves into the computer, see Kirschenbaum's forensic history of electronic writing and hard disks: Matthew Kirschenbaum, *Mechanisms: New Media and the Forensic Imagination* (Cambridge, MA: MIT Press, 2008).

8. Igo, *The Known Citizen*, 222.

9. Carolyn L. Geda, "Social Science Data Archives," *The American Archivist* 42, no. 2 (1979): 159.

10. Senate Committee on the Judiciary, Subcommittee on Constitutional Rights, "Federal Databanks and Constitutional Rights," 1971, iv.

11. During the 1940s, leading survey research centers were founded throughout the US, including the Bureau of Applied Social Research at Columbia University (1944), and the National Opinion Research Center at the University of Chicago (1941), and the Survey Research Center at University of Michigan (1946).

12. Ralph L. Bisco, "Social Science Data Archives: A Review of Developments," *The American Political Science Review* 60, no. 1 (1966): 93–109, https://doi.org/10.2307/1953810.

13. Loughnane and Aspray, "Rethinking the Call for a US National Data Center in the 1960s," 213.

14. Today, the data archive is known as the "Inter-University Consortium for Political and Social Research" and is made up of more than 750 academic research institutions.

15. Ralph L. Bisco, "Social Science Data Archives Technical Considerations," *Social Science Information* 4, no. 3 (September 1, 1965): 129–150, https://doi.org/10.1177/053901846500400311.

16. Charles M. Dollar, "Computers, the National Archives, and Researchers," *Prologue-Quarterly of the National Archives* 8, no. 1 (1976): 29–34.

17. David M. Pemberton, "'Blisters on My Heels, Corns on My Toes': Taking the 1930 Census of Population," *National Archives* 34, no. 4 (2002), https://www.archives.gov/publications/prologue/2002/winter/1930-census.html; Bouk, "The National Data Center and the Rise of the Data Double."

18. Columbia University Bureau of Applied Social Research, *Specialized Social Science Information Services in the United States / Jack Ferguson* (New York: Distributed by Clearinghouse for Federal Scientific and Technical Information, 1965).

19. Bisco, "Social Science Data Archives Technical Considerations," 147.

20. Carl Beck, "In Memoriam: Ralph L. Bisco," *PS: Political Science & Politics* 3, no. 3 (1970): 467–467, https://doi.org/10.1017/S1049096500029413.

21. Ralph L. Bisco and Council of Social Science Data Archives, *Data Bases, Computers, and the Social Sciences*, ed. Ralph L. Bisco, Information Sciences Series (New York: Wiley-Interscience, 1970).

22. Many of the earliest data archives for research in the US were funded through the newly established National Science Board, known as the "National Science Foundation" today.

23. Bisco, "Social Science Data Archives Technical Considerations," 131.

24. Today, this preparation labor is what archivists typically do in traditional collecting institutions—spending a fair amount of time creating descriptive metadata for materials so that they are easily accessible when researchers are searching, thus making them searchable in catalog or electronic resource databases.

25. Frieda B. Libaw, "Information Handling in the Behavioral Sciences: Report of a First Convocation of a Conglomerate Clan," *American Behavioral Scientist* 10, no. 6 (February 1, 1967): 8–12, https://doi.org/10.1177/000276426701000603.

26. Bisco, "Social Science Data Archives Technical Considerations," 144.

27. Maurice B. Line, "Information Requirements in the Social Sciences: Some Preliminary Considerations," *Journal of Librarianship* 1, no. 1 (January 1, 1969): 1–19, https://doi.org/10.1177/096100066900100101; Libaw, "Information Handling in the Behavioral Sciences."

28. Richard Ruggles et al., "Report of the Committee on the Preservation and Use of Economic Data (1965)," accessed November 16, 2021, http://archive.org/details/ReportOfTheCommitteeOnThePreservationAndUseOfEconomicData1965, 4.

29. Ruggles et al., "Report of the Committee on the Preservation and Use of Economic Data (1965), "19–20.

30. "Privacy and Efficient Government: Proposals for a National Data Center," *Harvard Law Review* 82, no. 2 (December 1968): 400, https://doi.org/10.2307/1339231.

31. Jack Sawyer and Howard Schechter, "Computers, Privacy, and the National Data Center: The Responsibility of Social Scientists," *American Psychologist* 23, no. 11 (November 1968): 812, https://doi.org/10.1037/h0026719.

32. Ruggles et al., "Report of the Committee on the Preservation and Use of Economic Data (1965)", 18.

33. However, in 1966, the Census Bureau began to plan for a data user program. See Kraus, "Statistical Déjà Vu," 5.

34. Edgar S. Dunn, "Review of Proposal for a National Data Center, Statistical Evaluation Report" (US Bureau of the Budget, Office of Statistical Standards, December 1965).

35. Carl Kaysen et al., "Report of the Task Force on the Storage of and Access to Government Statistics," *The American Statistician* 23, no. 3 (1969): 11–19.

36. Kaysen et al., "Report of the Task Force on the Storage of and Access to Government Statistics," 11.

37. Kaysen et al., "Report of the Task Force on the Storage of and Access to Government Statistics," 19.

38. Alan F. Westin, *Privacy and Freedom*, 1st ed. (New York: Atheneum, 1967), 317.

39. U.S. Senate Committee on the Judiciary, Subcommittee on Administrative Practice and Procedure, "Government Dossier (Survey of Information Contained in Government Files)," 90th Cong., 1st sess., 1967, 26–27.

40. Lepore, *If Then*, 281.

41. In the US, every major computer from the 1940s–1950s was funded by the government and usually tied to war efforts. Even the first commercially available computer was purchased by the Census Bureau. For more on this history, see James W. Cortada, *Before the Computer: IBM, NCR, Burroughs, & Remington Rand & the Industry They Created, 1865–1956* (Princeton, NJ: Princeton University Press, 2000).

42. Igo, *The Known Citizen*, 226.

43. U.S. House Committee on Government Operations, Special Subcommittee on Invasion of Privacy, "The Computer and Invasion of Privacy: Hearings," 89th Cong., July 26–28, 1966 (Washington, DC, 1966).

44. U.S. Senate Committee on the Judiciary, Subcommittee on Administrative Practice and Procedure, "Invasions of Privacy (Government Agencies): Hearings," 89th Cong., 2nd sess., pt. 5, (Washington, DC, 1966).

45. Morris S. Ogul, *Congress Oversees the Bureaucracy: Studies in Legislative Supervision / Morris S. Ogul* (Pittsburgh, PA: University of Pittsburgh Press, 1976), 114.

46. Priscilla M. Regan, *Legislating Privacy: Technology, Social Values, and Public Policy* (Chapel Hill: University of North Carolina Press, 1995), 72.

47. Edgar S. Dunn, "The Idea of a National Data Center and the Issue of Personal Privacy," *The American Statistician* 21, no. 1 (1967): 21, https://doi.org/10.2307/2681910.

48. Nan Robertson, "Data-Center Aims Scored in Inquiry: Advocates Hear Legislators Challenge Their System," *New York Times*, 1966.

49. Dunn, "The Idea of a National Data Center and the Issue of Personal Privacy," 23.

50. Or perhaps more cynically, actual experience with the government paperwork culture at the time.

51. Igo, *The Known Citizen*, 223.

52. Dunn, "The Idea of a National Data Center and the Issue of Personal Privacy," 21.

53. T. D. Woolsey, "Databanks Are Not the Answer: A Statistician's Viewpoint," *American Journal of Public Health and the Nation's Health* 60, no. 10 (October 1, 1970): 1991–1995, https://doi.org/10.2105/AJPH.60.10.1991.

54. James W. Ramey, "Urban Databanks and the Rights of the Individual," *Socio-Economic Planning Sciences* 1, no. 3 (July 1, 1968): 331, https://doi.org/10.1016/0038-0121(68)90019-0.

55. High-profile cases included the Army's secret files on anti-war protestors, previously unknown collection of dossiers on federally funded scientists kept by the National Science Foundation, and a database of immigrant children. Alan F. Westin and Michael A. Baker, *Databanks in a Free Society: Computers, Record-Keeping and Privacy*, New York, NY: Quadrangle Books, 1972.

56. Sawyer and Schechter, "Computers, Privacy, and the National Data Center," 114.

57. Executive Office of the President, Office of Science and Technology, "Privacy and Behavioral Research" (Washington, DC, February 1967).

58. T. J. Vander Noot, "Computers, Social Science Statisticians, and the A.S.A.," *The American Statistician* 22, no. 3 (1968): 21, https://doi.org/10.2307/2681798.

59. "The Computer and Invasion of Privacy: Hearings."

60. Cornelius Gallagher, "Privacy and the National Data Center," Speech presented at the Joint Computer Conference, Atlantic City, NJ, April 18, 1967.

61. Ralph L. Bisco, "Social Science Data Archives: Progress and Prospects," *Social Science Information* 6, no. 1 (February 1, 1967): 70, https://doi.org/10.1177/053901846700600102.

62. Sawyer and Schechter, "Computers, Privacy, and the National Data Center," 811.

63. For example, files on tape were stored serially and had to be read from the beginning to the end of a length of tape to retrieve specific information.

64. Raymond T. Bowman, "Crossroad Choices for the Future Development of the Federal Statistical System," *Journal of the American Statistical Association* 63, no. 323 (1968): 801–816, https://doi.org/10.2307/2283874.

65. Raymond T. Bowman, "The Idea of a Federal Statistical Data Center—It's Purposes and Structure," in *Data Bases, Computers, and the Social Sciences*, ed. Ralph L. Bisco, Information Sciences Series (New York: Wiley-Interscience, 1970), 66–67.

66. Luther J. Carter, "National Data Bank: Its Advocates Try to Erase 'Big Brother' Image," *Science* 163, no. 3863 (1969): 161.

67. Bowman, "The Idea of a Federal Statistical Data Center," 69.

68. Malcolm Warner and Michael Graham Stone, *The Data Bank Society: Organizations, Computers and Social Freedom* (London: Allen & Unwin, 1970), 20.

69. Carter, "National Databank," 160.

70. Westin was a professor of law at Columbia University, who had previously published *Privacy and Freedom*, which was considered a foundational text in US privacy law.

71. Westin and Baker, "Databanks in a Free Society," 30.

72. Kenneth Osgood et al., eds., "'We Need to Get a Better Story to the American People': LBJ, the Progress Campaign, and the Vietnam War on Television," in *Selling War in a Media Age: The Presidency and Public Opinion in the American Century* (Gainesville: University Press of Florida, 2010), https://doi.org/10.5744/florida/9780813034669.003.0008.

73. The journalist Murrey Marder is credited as having popularized the phrase during the Johnson administration in the Washington Post. "Credibility gap" has since come to mean the discrepancy between reality and accounts from government officials and politicians.

74. Lepore, *If Then*, 313.

75. Steven Lubar, "'Do Not Fold, Spindle or Mutilate': A Cultural History of the Punch Card," *Journal of American Culture* 15, no. 4 (1992): 43–55, https://doi.org/10.1111/j.1542-734X.1992.1504_43.x.

76. Students and university researchers would also march on and occupy at least thirteen campus computer centers from 1968 to 1972. Julyk (2008) Table 1, 214–215. David P. Julyk, "The Trouble With Machines Is People: The Computer as Icon in Post-War America: 1946-1970." PhD dissertation, The University of Michigan, 2008.

77. Fred Turner, *From Counterculture to Cyberculture: Stewart Brand, the Whole Earth Network, and the Rise of Digital Utopianism* (Chicago, IL: University of Chicago Press, 2006), 12.

78. More than 260 bills featuring privacy rights were introduced from 1965 to 1972, and more than fifty congressional hearings and reports explored the impact of new technology on privacy. See James B. Rule (1980). James B. Rule, *The Politics of privacy: planning for personal data systems as powerful technologies* (New York: Elsevier, 1980).

79. "Federal Data Banks, Computers and the Bill of Rights. Hearings before the Subcommittee on Constitutional Rights of the Committee on the Judiciary, Part 1," (Washington, DC, 1971); US Department of Health, Education and Welfare, Secretary's Advisory Committee on Automated Personal Data Systems, "Records, Computers, and the Rights of Citizens" (Washington, DC: Government Printing Office, 1973).

80. Arthur R. Miller, *The Assault on Privacy: Computers, Data Banks, and Dossiers* (Ann Arbor: University of Michigan Press, 1971), 56.

81. Westin and Baker, "Databanks in a Free Society," 20.

82. Miller, *The Assault on Privacy*, 59.

83. Igo, *The Known Citizen*, 224.

84. Miller, *The Assault on Privacy*, 67.

CHAPTER 3

1. Platform models to data access are discussed further in chapter 6.

2. Gordon Bell, "Computer Generations" (Digital Equipment Corporation, 1975), Box 60, Folder 4, Computer History Museum, Digital Equipment Corporation records.

3. John E. Yates, "A Time Sharing System for the PDP-1 Computer," Scientific Report No. 5, Electronic Systems Laboratory (Cambridge, MA: Massachusetts Institute of Technology, June 1962), 1, Box 4, Folder 1, Computer History Museum, PDP-1 records.

4. Yates, "A Time Sharing System for the PDP-1 Computer," 2.

5. DEC donated a PDP-1 to the electrical engineering department at MIT in 1961, where they began to experiment with time-sharing systems to make computer time more accessible to students. Jack B. Dennis, "A Time-Shared Computation Facility for Education and Research" (Massachusetts Institute of Technology, May 16, 1963), Box 3, Folder 8, Computer History Museum, PDP-1 records.

6. Ceruzzi, *A History of Modern Computing*, 125. Gordon Bell, "Toward a History of (Personal) Workstations," in *Proceedings of the ACM Conference on The History of Personal Workstations*, HPW '86 (New York: Association for Computing Machinery, 1986), 1–17, https://doi.org/10.1145/12178.12179.

7. Dennis, "A Time-Shared Computation Facility for Education and Research," 1–2.

8. Campbell-Kelly et al., *Computer*, 106–107.

9. For more on this distinction and the economic ideology of the user, see Hu (2015).

10. This chapter specifically focuses on DEC programmed data processor minicomputers. For a range of the most popular commercial minicomputers with integrated circuits manufactured from 1965–1974, see Ceruzzi, *A History of Modern Computing*, 192 (table 6.1).

11. C. Gordon Bell et al., *Computer Engineering: A DEC View of Hardware Systems Design* (Bedford, MA: Digital Press, 2014).

12. "Programmed Data Processor-1 Manual" (Digital Equipment Corporation, 1961), 102664957, Computer History Museum, https://www.computerhistory.org/pdp-1/50fc0d6d57d4a4d67885984808c8d847/.

13. Edward Fredkin, "The Time Sharing of Computers," *Computers and Automation*, November 1963.

14. In particular, PDPs used in Marvin Minsky's artificial intelligence lab at MIT, the TX-0 and TX-2 at Lincoln Laboratory (also located at MIT), as well as PDPs and IBM machines at Stanford's AI Lab. See, e.g., Lepore, *If Then*; Chun (2011), Campbell-Kelly et al. (2019), and Levy (2010), among others.

15. For more on data processing practices, see Thomas Haigh, "How Data Got Its Base: Information Storage Software in the 1950s and 1960s," *IEEE Annals of the History of Computing* 31, no. 4 (October 2009): 6–25, https://doi.org/10.1109/MAHC.2009.123.

16. For more indexicality in data structures, see Feinberg, *Everyday Adventures with Unruly Data* (2022).

17. In contemporary computing systems, Random Access Memory ("RAM") is memory and Read Only Memory ("ROM") retains data when the computer is shut down. Today, both RAM and ROM are enclosed in personal computers.

18. Campbell-Kelly et al., *Computer*, 100.

19. Data is recorded to magnetic tape by magnetizing spots in the ferric oxide coating of the tape. The ferric oxide tape can retain the magnetization as nonvolatile

storage. When a tape with recorded data moves across a tape drive head, its magnetization induces current in a coil, and represented data can be sensed so that the drive mechanism can "read" the tape.

20. DECtape's composition was unique in that it had a sandwich-style composition: a magnetic middle layer between two layers of mylar laminate. This was a major improvement over paper tape.

21. Marcia Blumenthal, "Fraud Ruled in 1968 DEC Tape Patent," *Computerworld*, May 1, 1978.

22. "DECTape Transport Design," *Electromechanical Design: Components & Systems*, November 1964, Box 179, Folder 12, Computer History Museum, Digital Equipment Corporation records, https://deramp.com/downloads/mfe_archive/011-Digital%20Equipment%20Corporation/20%20Magnetic%20Tape%20Subsystems/dectapeDesignNov64.pdf.

23. "DECTape Control 551 Preliminary Manual" (Digital Equipment Corporation, 1965), Box 179, Folder 14, Computer History Museum, Digital Equipment Corporation records, http://www.bitsavers.org/pdf/dec/dectape/551/H-551P_DECtape_Control_551_Preliminary.pdf.

24. "Microtape: 555 Tape Transport/ 550 Tape Control/ Programs" (Digital Equipment Corporation, 1964), Box 7, Folder 13, Digital Equipment Corporation records, http://www.computinghistory.org.uk/userdata/files/dec-microtape-555-tape-transport-550-tape-control-programs.pdf.

25. This innovative approach came from another dual tape drive system developed for the TX-2 computer created by Wesley Clark at the Lincoln Laboratory at MIT. For more, see Bell et al., *Computer Engineering*, 215; Mary Allen Wilkes and W. A. Clark, *Programming the LINC* (St. Louis, MO: Computer Research Laboratory, Washington University, 1965).

26. "550 DECTape Control Instruction Manual" (Digital Equipment Corporation, 1966), Box 4A, Folder 5, 102652390, Computer History Museum, PDP-1 records.

27. Digital Equipment Corporation, *PDP11: TC11 DECtape System Manual* (Maynard, MA: Digital Equipment Corporation, 1971).

28. "PDP-6 Time Sharing Software" (Digital Equipment Corporation, 1965), Box 60, Folder 5, Digital Equipment Corporation records, http://bitsavers.org/pdf/dec/pdp6/PDP-6_TimsharingBroch.pdf.

29. Fernando Corbató et al., "An Experimental Time-Sharing System," *Proceedings of the AIEE-IRE*, May 1, 1962, https://dl.acm.org/doi/abs/10.1145/1460833.1460871.

30. John McCarthy, "Reminiscences on the History of Time-Sharing," *IEEE Annals of the History of Computing* 14, no. 1 (January 1, 1992): 19–24.

31. James W. Cortada, *Historical Dictionary of Data Processing: Technology* (New York: Greenwood, 1987), 108.

32. Yates, "A Time Sharing System for the PDP-1 Computer," 3.

33. Thomas N. Pyke, "Time-Shared Computer Systems," *Advances in Computers* 8 (1967): 1–45, https://doi.org/10.1016/S0065-2458(08)60693-2.

34. C. Gordon Bell and Michael M. Gold, "An Introduction to the Structure of Time-Shared Computers," in *Advances in Information Systems Science: Volume 4*, ed. Julius T. Tou (Boston, MA: Springer US, 1972), 161–272, https://doi.org/10.1007/978-1-4615-9053-8_4.

35. For descriptions of "real-time interaction," see early brochures such as "PDP-6 Time Sharing Software."

36. Don Libes and Sanford Ressler, *Life with UNIX: A Guide for Everyone* (Englewood Cliffs, NJ: Prentice Hall, 1989).

37. D. M. Ritchie, "The UNIX System: The Evolution of the UNIX Time-Sharing System," *AT&T Bell Laboratories Technical Journal* 63, no. 8 (1984): 1578, https://doi.org/10.1002/j.1538-7305.1984.tb00054.x.

38. Berkeley Software Distribution Records, CU-551, The Bancroft Library, University of California, Berkeley.

39. The RK05 was an early DEC disk cartridge drive common for PDP-11s. Andrew Tannenbaum, "Politics of UNIX," Washington, DC USENIX Conference, 1984. As quoted in Libes and Ressler, *Life with Unix*, 13.

40. Peter Salus, *A Quarter Century of UNIX* (Reading, MA: Addison-Wesley, 1994), 138–39, https://dl.acm.org/doi/abs/10.5555/191771.

41. Salus, *A Quarter Century of UNIX*, 155.

42. Salus, *A Quarter Century of UNIX*, 143.

43. Christopher M. Kelty, *Two Bits: The Cultural Significance of Free Software*, Experimental Futures (Durham, NC: Duke University Press, 2008), 131.

44. Bradley Fidler and Amelia Acker, "Metadata, Infrastructure, and Computer-Mediated Communication in Historical Perspective," *Journal of the Association for Information Science and Technology* 68, no. 2 (February 1, 2017): 412–422, https://doi.org/10.1002/asi.23660.

45. Daniel Siewiorek et al., *Computer Structures: Principles and Examples* (New York: McGraw Hill, 1982), https://dl.acm.org/doi/abs/10.5555/578056.

46. Warren Toomey, "The Strange Birth and Long Life of Unix," *IEEE Spectrum*, November 28, 2011, https://spectrum.ieee.org/the-strange-birth-and-long-life-of-unix.

47. Kelty, *Two Bits*, 130.

48. For background and interviews on reporting on the antagonism between DEC and Unix users, see Salus, *A Quarter Century of Unix*, 181–188.

49. Laine Nooney, Kevin Driscoll, and Kera Allen, "From Programming to Products: Softalk Magazine and the Rise of the Personal Computer User," *Information & Culture* 55, no. 2 (2020): 105–129; Atsushi Akera, "Voluntarism and the Fruits of Collaboration: The IBM User Group, Share," *Technology and Culture* 42, no. 4 (2001): 710–736.

50. Amelia Acker, "Emulation Practices for Software Preservation in Libraries, Archives, and Museums," *Journal of the Association for Information Science and Technology* 72, no. 9 (2021): 1148–1160, https://doi.org/10.1002/asi.24482; Amelia Acker, "Accessing Software: Emulation in Information Institutions," *Information & Culture* 59, no. 1 (2023).

51. Iron Mountain, "Tape Backup Advantages for Ransomware Attacks," *Iron Mountain*, August 2021, https://www.ironmountain.com/resources/general-articles/t/tape-backup-advantages-for-ransomware-attacks.

52. H. C. Hinz, "Magnetic Tape Technology in the 1990s," in *[1990] Digest of Papers. Tenth IEEE Symposium on Mass Storage Systems@m_Crisis in Mass Storage*, Monterey, CA (New York: IEEE, 1990), 38–42, https://doi.org/10.1109/MASS.1990.113566; J. C. Mallinson, "Magnetic Tape Recording: Archival Considerations," in *[1990] Digest of Papers. Tenth IEEE Symposium on Mass Storage Systems@m_Crisis in Mass Storage*, Monterey, CA (New York: IEEE, 1990), 58–61, https://doi.org/10.1109/MASS.1990.113570.

53. "PDP-42-2: Preliminary Microtape File System," PDP-1 Computer (Cambridge, MA: Electrical Engineering Department, Massachusetts Institute of Technology, February 6, 1970), Digital Equipment Corporation records.

54. Perhaps the most famous (and earliest) proclamation about the democratization of interactive computing was made by Stewart Brand in a 1972 *Rolling Stone* article about the time-sharing game called *Spacewar*: "Ready or not, computers are coming to the people. That's good news, maybe the best since psychedelics."

55. For more on the history of the shift from communal computing to individual computing experiences, see Rankin, *A People's History of Computing*. Rankin marks the academic time-sharing networks of the 1970s as early examples of social networking and personal computing before personal computers, but by the early 1980s, in her words, "the people computing became the people consuming" (Rankin, *A People's History of Computing*, 241).

CHAPTER 4

1. Campbell-Kelly et al., *Computer*; Haigh, "How Data Got Its Base"; Cortada, *Historical Dictionary of Data Processing*.

2. Edward V. Long, *The Intruders; the Invasion of Privacy by Government and Industry, by Edward V. Long. With a Foreword by Hubert H. Humphrey* (New York: Praeger, 1967).

3. Josh Lauer, *Creditworthy: A History of Consumer Surveillance and Financial Identity in America* (New York: Columbia University Press, 2017).

4. Laine Nooney, *The Apple II Age: How the Computer Became Personal* (Chicago, IL: University of Chicago Press, 2023).

5. Driscoll, *The Modem World*, 91.

6. For more on the cultural history of early networking and filesharing practices, please see Driscoll's *Modem World*, 90–131.

7. A. Christopher, "Wherefore Art Thou, Apple?" *Upside*, October 1991; J. Jerney, "Talking with Palm Chair Jeff Hawkins," *Pen Computing*, January 1995.

8. As containers, files go back centuries across many types of information systems, in what Cornelia Vismann calls the "epoch of files," stretching from chanceries to government dossiers. Cornelia Vismann, *Files: Law and Media Technology*, trans. Geoffrey Winthrop-Young (Stanford, CA: Stanford University Press, 2008), 12.

9. Zoë Sofia, "Container Technologies," *Hypatia* 15, no. 2 (April 2000): 181–201, https://doi.org/10.1111/j.1527-2001.2000.tb00322.x.

10. For more on how automated recordkeeping and computerized workplaces impacted the role and practices of archivists in preservation planning, see Anne J. Gilliland, "Reconceptualizing Records, the Archive and Archival Roles and Requirements in a Networked Society," *Book Science* 63, no. 63 (January 1, 2014): 17–34, https://doi.org/10.15388/Knygotyra.2014.63.4011; Meyer H. Fishbein, "Automation in Archives: A Summary History," National Archives, Maryland, 1981; Kenneth Thibodeau, "Overview of Technological Approaches to Digital Preservation and Challenges in Coming Years" (Washington, DC: Council on Library and Information Resources, 1997).

11. The stack is discussed more in depth in the next chapter on metadata and wireless networking.

12. It is worth noting, as others have, that early designs for portable devices assumed larger hands and men's front shirt pockets.

13. Craig Robertson, *The Filing Cabinet: A Vertical History of Information* (Minneapolis: University of Minnesota Press, 2021).

14. Many human–computer interaction researchers have found since the 1980s that digital filing systems and the practices people build up around them largely replicate experiences and filing strategies of analog paper environments using filing cabinets, working files, and inboxes and outboxes on desks. See Abigail J. Sellen and Richard Harper, *The Myth of the Paperless Office* (Cambridge, MA: MIT Press,

2002); Ofer Bergman and Steve Whittaker, *The Science of Managing Our Digital Stuff* (Cambridge, MA: MIT Press, 2016).

15. Richard Harper et al., "What Is a File?," in *Proceedings of the 2013 Conference on Computer Supported Cooperative Work*, CSCW '13 (New York: Association for Computing Machinery, 2013), 1125–1136, https://doi.org/10.1145/2441776.2441903.

16. Day, *Indexing It All*.

17. Melanie Feinberg, "Reading Databases: Slow Information Interactions beyond the Retrieval Paradigm," *Journal of Documentation* 73, no. 2 (February 22, 2017): 336–356, https://doi.org/10.1108/JD-03-2016-0030.

18. David E. Karger, "Unify Everything: It's All the Same to Me," in *Personal Information Management*, ed. William P. Jones and Jaime Teevan (Seattle: University of Washington Press, 2007), 137, http://site.ebrary.com/id/10448623.

19. John Tang et al., "What Do You See in the Cloud? Understanding the Cloud-Based User Experience through Practices," Microsoft, September 2, 2013, https://www.microsoft.com/en-us/research/publication/what-do-you-see-in-the-cloud-understanding-the-cloud-based-user-experience-through-practices/.

20. Casio was the original equipment manufacturer for Tandy hardware, and both companies released versions of Zoomer at the same time. Butter and Pogue, *Piloting Palm*, 22.

21. Rashi Glazer et al., *Palm Computing in 1994* (London: Sage, 2000), 9, https://doi.org/10.4135/9781526410788.

22. "Developer Forum: Interview with Jeff Hawkins," *PDA News*, March 1994, 5, Box 1, Folder 1, Computer History Museum, Jeff Hawkins papers.

23. Dave Andrews, "Personal Digital Assistants: Behind the Wheel of the First Zoomer and Newton PDAs," *BYTE*, September 1, 1993.

24. Anonymous, "Portable and Personal: Take Your Office with You," *Managing Office Technology* 38, no. 11 (November 1993): 49.

25. When it was released, the Doonesbury comic strip lampooned it for a full week. Executives claimed that Apple's colossal nosedive had sunk all other PDA releases.

26. *Computer Chronicles 1993 PDA Review Part 3 of 5*, 2009, https://www.youtube.com/watch?v=BPt-SOtnyE0.

27. Much of the slowness in early digital assistants and tablets was from by the GEOS operating system. For more on how Palm designers incorporated a set of user interface design principles focused optimizing a faster UI, see PalmSource, *Zen of Palm: Designing Products for Palm OS* (Sunnyvale, CA: PalmSource, 2003), http://archive.org/details/zen-of-palm.

28. These expandable modem ports allowed Zoomers to connect telecom networks.

29. Danny Atkin, "Casio Z-7000; Tandy Z-PDA. (Personal Digital Assistants) Hardware Review," *COMPUTE!* March 1994.

30. "Zoomer Package Both Functional and Promising," *PDA News*, March 1994, 9, Box 1, Folder 1, Computer History Museum, Jeff Hawkins papers.

31. "Zoomer Package Both Functional and Promising"; Andrews, "Personal Digital Assistants" 1993.

32. "Simon-Ized: Part Cellular Phone, Part Touch Sensitive Key Pad, BellSouth's Simon May Revolutionize the Industry," *PDA News*, March 1994, Vol. 1, Number 3 edition.

33. PalmSource, *Zen of Palm*.

34. "The Palm Personal Information Manager" (Palm, 1993), 1, Box 1, Folder 1, Computer History Museum, Jeff Hawkins papers.

35. "Palm Position Paper" (Palm Computing Inc., 1993), Box 1, Folder 1, Computer History Museum, Jeff Hawkins papers.

36. Glazer et al., *Palm Computing in 1994*.

37. Butter and Pogue, *Piloting Palm*, 57.

38. Much of the early software development for PDA software would foreshadow the app store models found in contemporary iOS and Android environments, which is discussed in detail in chapter 6.

39. G. Christian Hill and Jim Carlton, "Telecommunications (A Special Report): A Wireless World—Getting Personal: The Newton Has Given Personal Digital Assistants a Bad Name; But Don't Be Fooled: PDAs Will Soon Revolutionize the Way People Communicate," *Wall Street Journal*, Eastern Edition, February 11, 1994.

40. Agar, *Constant Touch*.

41. Karlin Lillington, "PayPal Puts Dough in Your Palm," *Wired*, July 27, 1999, https://www.wired.com/1999/07/paypal-puts-dough-in-your-palm/.

42. Butter and Pogue, *Piloting Palm*, 57.

43. Agar, *Constant Touch*.

44. Paul Eng and Amy Dunkin, "Smart, Useful—and They Won't Put a Sag in Your Suit," *BusinessWeek*, no. 3374 (May 30, 1994): 141–142.

45. NB. The first edition of the book uses the word "monad," but the meaning appears to be "nomad." See https://www.google.com/books/edition/Wirelessness/0Q9gJX-WjLMC?hl=en&gbpv=1&bsq=a%20troubled%20figure,%20plunging%20and%20ascending%20between%20freedom%20and%20control,%20sometimes%20configured%20as%20windowless.

46. Wendy Hui Kyong Chun, *Control and Freedom: Power and Paranoia in the Age of Fiber Optics* (Cambridge, MA: MIT Press, 2006), 249.

47. By the early 2000s, more than 65 percent of Americans owned cell phones, and by 2010, more than 35 percent owned smartphones. The attendant "apps culture" emerging because of smartphone use for games and shopping was documented by Pew Research and Nielson in the early 2010s. See Nielsen Research, "The State of Mobile Apps," *Nielsen*, June 2010, https://www.nielsen.com/insights/2010/the-state-of-mobile-apps-2/; Kristen Purcell et al., "The Rise of Apps Culture," *Pew Research Center: Internet, Science & Tech* (blog), September 14, 2010, https://www.pewresearch.org/internet/2010/09/14/the-rise-of-apps-culture/; John Horrigan, "Consumption of Information Goods and Services in the U.S.," *Pew Research Center: Internet, Science & Tech* (blog), November 23, 2003, https://www.pewresearch.org/internet/2003/11/23/consumption-of-information-goods-and-services-in-the-u-s/.

48. Tang et al., "What Do You See in The Cloud?"; Krishna Akhil Kumar Adavi and Amelia Acker, "What Is a File on a Phone? Personal Information Management Practices Amongst WhatsApp Users," *Proceedings of ACM Human–Computer Interaction 7, CSCW2, Article 372*, 2023, 1–28, https://doi.org/10.1145/3610221.

49. Catherine C. Marshall, "How People Manage Personal Information over a Lifetime," in *Personal Information Management*, ed. William P. Jones and Jaime Teevan (Seattle: University of Washington Press, 2011), 57–75.

50. For a survey, see chapter 4 in Oscar H. Gandy Jr., *The Panoptic Sort: A Political Economy of Personal Information* (Boulder, CO: Westview Press, 1993).

CHAPTER 5

1. Michael K. Buckland, "Information as Thing," *Journal of the American Society for Information Science* 42, no. 5 (June 1, 1991): 351–360, https://doi.org/10.1002/(SICI)1097-4571(199106)42:5<351:AID-ASI5>3.0.CO;2-3.

2. Sebastian Benthall et al., "Contextual Integrity Through the Lens of Computer Science," *Foundations and Trends in Privacy and Security* 2, no. 1 (December 22, 2017): 1–69, https://doi.org/10.1561/3300000016.

3. Rob Guay and Kean Birch, "A Comparative Analysis of Data Governance: Socio-Technical Imaginaries of Digital Personal Data in the USA and EU (2008–2016)," *Big Data & Society* 9, no. 2 (July 1, 2022): 20539517221112925, https://doi.org/10.1177/20539517221112925.

4. National Security Agency/Central Security Service, "NSA Reports Data Deletion," June 28, 2018, https://www.nsa.gov/Press-Room/Press-Releases-Statements/Press-Release-View/Article/1618691/nsa-reports-data-deletion/http%3A%2F%2Fwww.nsa.gov%2FPress-Room%2FPress-Releases-Statements%2FPress-Release-View%2FArticle%2F1618691%2Fnsa-reports-data-deletion%2F.

5. Adam I. Klein et al., "Privacy and Civil Liberties Oversight Board Report on the Government's Use of the Call Detail Records Program Under the USA Freedom Act," February 2020, 106.

6. Ellen Nakashima, "NSA Has Halted a Counterterrorism Program Relying on Phone Records amid Doubts about Its Utility," *Washington Post*, March 6, 2019, https://www.washingtonpost.com/world/national-security/nsa-has-halted-a-counterterrorism-program-relying-on-phone-records-amid-doubts-about-its-utility/2019/03/05/f2d2793e-3f80-11e9-922c-64d6b7840b82_story.html.

7. Office of Director of National Intelligence, "Statistical Transparency Report Regarding the Use of National Security Authorities: Calendar Year 2018," Office of Civil Liberties, Privacy, and Transparency, 2019; Office of Director of National Intelligence, "Statistical Transparency Report Regarding the Use of National Security Authorities: Calendar Year 2017," Office of Civil Liberties, Privacy, and Transparency, 2018; NSA Civil Liberties and Privacy Office, "Transparency Report: THE USA FREEDOM Act Business Records FISA Implementation," January 15, 2016.

8. Dustin Volz and Warren P. Strobel, "NSA Recommends Dropping Phone-Surveillance Program," *Wall Street Journal*, accessed December 11, 2022, https://www.wsj.com/articles/nsa-recommends-dropping-phone-surveillance-program-11556138247.

9. "Text—H.R.2048—114th Congress (2015–2016): USA FREEDOM Act of 2015," June 2, 2015, http://www.congress.gov/.

10. Glenn Greenwald, "NSA Collecting Phone Records of Millions of Verizon Customers Daily," *The Guardian*, June 6, 2013, sec. US news, https://www.theguardian.com/world/2013/jun/06/nsa-phone-records-verizon-court-order.

11. Previously, the USA Patriot Act allowed collecting up to four hops from a target's CDRs. This would include both known contacts and incoming calls, including telemarketing calls, spoofing, and robo calls.

12. Two hops would be two degrees from the initial person of interest's CDRs—that is, all the call records from people contacted by the target, and all the call records from those who were contacted by the people contacted by the target.

13. For more on how metadata can be data, see Matthew S. Mayernik, "Metadata," *Knowledge Organization* 47, no. 8 (2020): 696–713, https://doi.org/10.5771/0943-7444-2020-8-696; Matthew S. Mayernik, "Metadata Accounts: Achieving Data and Evidence in Scientific Research," *Social Studies of Science* 49, no. 5 (October 1, 2019): 732–757, https://doi.org/10.1177/0306312719863494.

14. Many information scholars have discussed the power of naming in data-driven systems such as classification, cataloging, or publishing linked open data. See, for example, María A. Matienzo, "To Hell with Good Intentions: Linked Data, Community and the Power to Name," *LITA Forum*, 2015, https://scholar.google.com/scholar?cluster=10216571721061765160&hl=en&oi=scholarr; Hope A. Olson, *The Power to Name: Locating the Limits of Subject Representation in Libraries* (Dordrecht: Springer, 2011); Amelia Acker, "Metadata," in *Uncertain Archives*, ed. Nanna Bonde

Thylstrup, Daniela Agostinho, Annie Ring, Catherine D'Ignazio, and Kristin Veel (Cambridge, MA: MIT Press, 2021), 321–330, https://doi.org/10.7551/mitpress/12236.003.0035.

15. Fidler and Acker, "Metadata, Infrastructure, and Computer-Mediated Communication in Historical Perspective."

16. Amelia Acker, "The Short Message Service: Standards, Infrastructure and Innovation," *Telematics and Informatics* 31, no. 4 (November 1, 2014): 559–568, https://doi.org/10.1016/j.tele.2014.01.004.

17. Vincent D. Blondel et al., "A Survey of Results on Mobile Phone Datasets Analysis," *EPJ Data Science* 4, no. 1 (December 2015): 1–55, https://doi.org/10.1140/epjds/s13688-015-0046-0.

18. Amy Wesolowski et al., "Commentary: Containing the Ebola Outbreak—The Potential and Challenge of Mobile Network Data," *PLoS Currents* 6 (September 29, 2014): ecurrents.outbreaks.0177e7fcf52217b8b634376e2f3efc5e, https://doi.org/10.1371/currents.outbreaks.0177e7fcf52217b8b634376e2f3efc5e.

19. Corinna Cortes et al., "Computational Methods for Dynamic Graphs," *Journal of Computational and Graphical Statistics* 12, no. 4 (December 1, 2003): 950–970, https://doi.org/10.1198/1061860032742.

20. Linus Bengtsson et al., "Improved Response to Disasters and Outbreaks by Tracking Population Movements with Mobile Phone Network Data: A Post-Earthquake Geospatial Study in Haiti," *PLOS Medicine* 8, no. 8 (August 30, 2011): e1001083, https://doi.org/10.1371/journal.pmed.1001083.

21. White House, "Statement by the President," Whitehouse.gov, June 7, 2013, https://obamawhitehouse.archives.gov/the-press-office/2013/06/07/statement-president.

22. Alan Rusbridger, "The Snowden Leaks and the Public," *New York Review of Books*, November 21, 2013.

23. For more on the politics of formats and formatting, see Nathaniel Tkacz, *Being with Data: The Dashboarding of Everyday Life*, 1st ed. (Cambridge: Polity, 2022); Jonathan Sterne, *MP3: The Meaning of a Format* (Durham, NC: Duke University Press Books, 2012).

24. Kitchin, *The Data Revolution*; van Dijck, "Datafication, Dataism and Dataveillance."

25. Kitchin, *The Data Revolution*.

26. This is, however, changing with data protection regulation outside of the US, such as the EU's GDPR and India's Digital Personal Data Protection Bill.

27. Anne J. Gilliland, "Reflections on the Value of Metadata Archaeology for Recordkeeping in a Global, Digital World," *Journal of the Society of Archivists* 32, no. 1 (April 1, 2011): 103–118, https://doi.org/10.1080/00379816.2011.563934.

28. Indeed, a recently declassified report from the Office of the Director of National Intelligence shows that many US and foreign intelligence agencies have begun exploring purchasing Americans' personal data as "commercially available information" for intelligence services just as the private sector does. See Senior Advisory Group Panel, "Declassified Report on Commercially Available Information," Office of the Director of National Intelligence. Washington, DC: United States Office of the Director of National Intelligence, January 27, 2022, https://www.dni.gov/index.php/newsroom/reports-publications/reports-publications-2023/3700-odni-senior-advisory-group-panel-declassified-report-on-commercially-available-information.

29. Acker and Donovan, "Data Craft: A Theory/Methods Package for Critical Internet Studies," *Information, Communication & Society* 22, no. 11 (July 20, 2019): 1590–1609, https://doi.org/10.1080/1369118X.2019.1645194"; Acker, "Data Craft: The Manipulation of Social Media Metadata," Data & Society, New York, November 5, 2018, https://datasociety.net/output/data-craft/

30. The relational qualities of data have been extensively theorized in science studies and critical legal studies. For more, see Sabina Leonelli, "What Counts as Scientific Data? A Relational Framework," *Philosophy of Science* 82, no. 5 (December 2015): 810–821, https://doi.org/10.1086/684083; Salomé Viljoen, "A Relational Theory of Data Governance," *The Yale Law Journal* 131, no. 2 (November 2021): 573–654.

31. Anne J. Gilliland, "Setting the Stage," in *Introduction to Metadata*, ed. Murtha Baca, 3rd ed. (Los Angeles, CA: J. Paul Getty Trust, 2016), 18.

32. For more on US privacy legislation in transmitting message content, see Regan, *Legislating Privacy*, 109–143.

33. Day, *Indexing It All*, 29.

34. Borgman, *Big Data, Little Data, No Data: Scholarship in the Networked World* (Cambridge, MA: MIT Press, 2015).

35. Buckland, "Description and Search: Metadata as Infrastructure," *Brazilian Journal of Information Science: Research Trends* 0, no. 0 (March 20, 2007), https://doi.org/10.36311/1981-1640.2006.V0N0.02.P3.

36. Acker, "Data Craft"; Acker and Donovan, "Data Craft"; Acker and Flamm, "COVID-19 Community Archives and the Platformization of Digital Cultural Memory," in *Proceedings of the 54th Hawaii International Conference on System Sciences* (Atlanta, GA: Association for Information Systems, 2021), 2544–2553, http://hdl.handle.net/10125/70926.

37. Mayernik, "Metadata Accounts."

38. National Research Council et al., *Bulk Collection of Signals Intelligence: Technical Options* (Washington, DC: National Academies Press, 2015).

39. Kris, "The NSA and the USA Freedom Act," Lawfare, July 2, 2018, https://www.lawfareblog.com/nsa-and-usa-freedom-act.

40. Susan Landau, "Transactional Information Is Remarkably Revelatory," *Proceedings of the National Academy of Sciences* 113, no. 20 (2016): 5468.

41. National Research Council et al., *Bulk Collection of Signals Intelligence.*

42. Rita Raley, "Dataveillance and Countervailance," in *"Raw Data" Is an Oxymoron*, ed. Lisa Gitelman (Cambridge, MA: MIT Press, 2013), 123.

43. Susan Landau and Asaf Lubin, "Examining the Anomalies, Explaining the Value: Should the USA FREEDOM Act's Metadata Program Be Extended?" *Harvard National Security Journal* 11 (2020): 308-358.

44. Daniel Banoub and Sarah J. Martin, "Storing Value: The Infrastructural Ecologies of Commodity Storage," *Environment and Planning D: Society and Space* 38, no. 6 (December 1, 2020): 1101–1019, https://doi.org/10.1177/0263775820911942.

45. Claudio Ciborra, *The Labyrinths of Information: Challenging the Wisdom of Systems*, 1st ed. (New York: Oxford University Press, 2004).

CHAPTER 6

1. For more on the impact of wireless networks on culture, see José van Dijck, *The Culture of Connectivity: A Critical History of Social Media*, 1st ed. (Oxford: Oxford University Press, 2013); Adrian Mackenzie, *Wirelessness: Radical Empiricism in Network Cultures* (Cambridge, MA: MIT Press, 2010).

2. José van Dijck, Thomas Poell, and Martijn de Waal, *The Platform Society: Public Values in a Connective World* (New York, NY: Oxford University Press, 2018).

3. Steven J. Jackson et al., "Understanding Infrastructure: History, Heuristics, and Cyberinfrastructure Policy," *First Monday* 12, no. 6 (2007): 1–9.

4. Srnicek, *Platform Capitalism*; van Dijck et al., *The Platform Society*.

5. Tarleton Gillespie, "The Politics of 'Platforms,'" *New Media & Society* 12, no. 3 (May 1, 2010): 347–364, https://doi.org/10.1177/1461444809342738; Nanna Bonde Thylstrup, *The Politics of Mass Digitization* (Cambridge, MA: MIT Press, 2019).

6. Shilton, "Values Levers."

7. Michael Schudson, *The Rise of the Right to Know: Politics and the Culture of Transparency, 1945–1975* (Cambridge, MA: Belknap Press, 2018); Daniel Greene, *The Promise of Access* (Cambridge, MA: MIT Press, 2021), https://mitpress.mit.edu/9780262542333/the-promise-of-access/; Robert H. Wiebe, *The Search for Order, 1877–1920*, 1st ed. (New York: Hill and Wang, 1966).

8. Thomas Kelly, *Early Public Libraries: A History of Public Libraries in Great Britain Before 1850* (London: Library Association, 1966).

9. Joseph Leroy Harrison, "The Public Library Movement in the United States," *The Library* 1–8, no. 1 (January 1, 1896): 141–149, https://doi.org/10.1093/library/s1-8.1.141.

10. National Research Council, *Engaging Privacy and Information Technology in a Digital Age*, ed. James Waldo, Herbert Lin, and Lynette I Millett (Washington, DC: National Academies Press, 2007), 234.

11. For information institutions in the United States, the freedom of access to information has historically been at the center of professional codes of ethics in the field of librarianship, archives, and education. The American Library Association defines access as when there is "creation, distribution, and possession of information that is free from interference and censorship." American Library Association, "Our Statement on Access," American Library Association, September 26, 2016, http://www.ala.org/rusa/strategic-priorities/access.

12. American Library Association, "Privacy: An Interpretation of the Library Bill of Rights," American Library Association, July 7, 2006, http://www.ala.org/advocacy/intfreedom/librarybill/interpretations/privacy.

13. Marijel (Maggie) Melo et al., "Makerspaces in Libraries at U.S. Public Colleges and Universities: A Census," *Portal: Libraries and the Academy* 23, no. 1 (2023): 35–43, https://doi.org/10.1353/pla.2023.0007; Noah Lenstra, "The Library as Playground: How Games and Play Are Reshaping Public Culture," *Journal of Education for Library and Information Science* 64, no. 1 (January 2023): 102–103, https://doi.org/10.3138/jelis.2022-0041.

14. For a range of interpretations (including oppositional critiques and poignant considerations) on library and archival collections as data, see more than fifty position statements from the Collections as Data project, funded by the Mellon Foundation. Sally Chambers et al., "Position Statements -> Collections as Data: State of the Field and Future Directions," May 2, 2023, https://doi.org/10.5281/zenodo.7897735.

15. For more on social movements and grass roots activism, see Kevin Driscoll, "From Punched Cards to 'Big Data': A Social History of Database Populism," *Communication +1*, no. 1 (August 29, 2012): 4.

16. Philip Mirowski, "The Future(s) of Open Science," *Social Studies of Science* 48, no. 2 (April 1, 2018): 171–203, https://doi.org/10.1177/0306312718772086.

17. Packingham v. North Carolina, No. 15–1194 (Supreme Court of the United States, June 19, 2017).

18. Adam Kriesberg et al., "An Analysis of Federal Policy on Public Access to Scientific Research Data," *Data Science Journal* 16 (June 14, 2017): 27, https://doi.org/10.5334/dsj-2017-027.

19. Office of the Press Secretary, "Fact Sheet: Data by the People, for the People—Eight Years of Progress Opening Government Data to Spur Innovation, Opportunity, & Economic Growth," Whitehouse.gov, September 28, 2016, https://obamawhitehouse.archives.gov/the-press-office/2016/09/28/fact-sheet-data-people-people-eight-years-progress-opening-government.

20. Many of the early members that led the USDS notably came from Google and Facebook. Some returned after serving during the Obama administration. For more, see Steven Levy, "The Final Days of Obama's Tech Surge," *Wired*, January 18, 2017, https://www.wired.com/2017/01/the-final-days-of-obamas-tech-surge/.

21. As part of the Obama administration's digital transition, I was granted early research access to the social media data. For more on the project, see Amelia Acker and Adam Kriesberg, "Tweets May Be Archived: Civic Engagement, Digital Preservation and Obama White House Social Media Data," *Proceedings of the Association for Information Science and Technology* 54, no. 1 (January 1, 2017): 1–9, https://doi.org/10.1002/pra2.2017.14505401001.

22. Bill Chappell, "Obama: Cyberspace Is the New 'Wild West,'" *NPR*, February 13, 2015, https://www.npr.org/sections/thetwo-way/2015/02/13/385960693/obama-to-urge-companies-to-share-data-on-cyber-threats.

23. SDKs and APIs are software that allow users (usually software developers) to access platform data.

24. Increasingly, we see different values emerging among developer cultures in mobile application development because of platforms' role in mediating access to data. See, for example, Rivka Ribak, "Translating Privacy: Developer Cultures in the Global World of Practice," *Information, Communication & Society* 22, no. 6 (May 12, 2019): 838–853, https://doi.org/10.1080/1369118X.2019.1577475; Min Ho Ryu et al., "Factors Affecting Application Developers' Loyalty to Mobile Platforms," *Computers in Human Behavior* 40 (November 1, 2014): 78–85, https://doi.org/10.1016/j.chb.2014.08.001; Katie Shilton and Daniel Greene, "Linking Platforms, Practices, and Developer Ethics: Levers for Privacy Discourse in Mobile Application Development," *Journal of Business Ethics* 155, no. 1 (March 1, 2019): 131–146, https://doi.org/10.1007/s10551-017-3504-8.

25. Amelia Acker, "A Death in the Timeline: Memory and Metadata in Social Platforms," *Journal of Critical Library and Information Studies* 2, no. 1 (2018): 27; Frank Pasquale, *The Black Box Society: The Secret Algorithms That Control Money and Information*, reprint ed. (Cambridge, MA: Harvard University Press, 2016).

26. Amelia Acker and Adam Kriesberg, "Social Media Data Archives in an API-Driven World," *Archival Science* 20 (2020: 105–123, https://doi.org/10.1007/s10502-019-09325-9; Deen Freelon, "Computational Research in the Post-API Age," *Political Communication* 35, no. 4 (October 2, 2018): 665–668, https://doi.org/10.1080/10584609.2018.1477506.

27. Rebekah Tromble, "Where Have All the Data Gone? A Critical Reflection on Academic Digital Research in the Post-API Age," *Social Media + Society* 7, no. 1 (January 1, 2021): 2056305121988929, https://doi.org/10.1177/2056305121988929.

28. Kate Klonick, “The New Governors: The People, Rules, and Processes Governing Online Speech,” *Harvard Law Review* 131, no. 6 (2018): 1598–1670.

29. Jon Porter, “Twitter Announces New API Pricing, Posing a Challenge for Small Developers,” *The Verge*, March 30, 2023, https://www.theverge.com/2023/3/30/23662832/twitter-api-tiers-free-bot-novelty-accounts-basic-enterprice-monthly-price.

30. At the time of writing, Twitter has recently been acquired by Elon Musk and is now known as “X.”

31. Michael Zimmer, “The Twitter Archive at the Library of Congress: Challenges for Information Practice and Information Policy,” *First Monday* 20, no. 7 (June 21, 2015), http://firstmonday.org/ojs/index.php/fm/article/view/5619; Elisabeth Fondren and McCune Meghan Menard, “Archiving and Preserving Social Media at the Library of Congress: Institutional and Cultural Challenges to Build a Twitter Archive,” *Preservation, Digital Technology & Culture* 47, no. 2 (2018): 33–44, https://doi.org/10.1515/pdtc-2018-0011.

32. Good faith research has found that the Library of Congress’s current approach to preserving published materials considers each work to be copyrighted material, and the burden of providing access to every single tweet as a copyrighted publication is too great for the corpus. Critics have speculated that the preservation policy to migrate and preserve data to particular formats (such as microfiche) is too prohibitive for providing research access when compared to the API means that researchers are now accustomed to because of platform APIs.

33. Chapter 4 discussed how Palm designed PalmConnect for developers to develop customized PDA apps that would connect mobile devices to information services like mobile payments with the early PayPal PDA app.

34. One of the most heavily open API models is Wikidata, managed by the Wikimedia foundation and leveraged by most search engines for knowledge panels and AI voice assistants. Another well-known one is the DPLA API, which has a distinct philosophy on API access. Digital Public Library of America, “Philosophy—Why Build the DPLA API?,” Digital Public Library of America, 2019, https://pro.dp.la/developers/philosophy.

35. Adam Kriesberg and Amelia Acker, “The Second US Presidential Social Media Transition: How Private Platforms Impact the Digital Preservation of Public Records,” *Journal of the Association for Information Science and Technology* 73, no. 11 (2022): 1529–1542, https://doi.org/10.1002/asi.24659.

36. David S. Ferriero, “Letter to Maloney on Trump Presidential Records on Social Media Platforms,” February 18, 2022, https://www.archives.gov/files/foia/ferriero-letter-to-maloney-on-trump-presidential-records-on-social-media-platforms.02.18.2022.pdf.

37. Although NARA was unable to collect complete account information, a special counsel investigation requested all internal data and records related to President Trump's account. Jaclyn Diaz, "Special Counsel Jack Smith Got a Secret Search Warrant for Trump's Twitter Account," *NPR*, August 9, 2023, sec. Law, https://www.npr.org/2023/08/09/1192953166/prosecutors-search-warrant-trump-twitter-account.

38. Nicholas A. John and Asaf Nissenbaum, "An Agnotological Analysis of APIs: Or, Disconnectivity and the Ideological Limits of Our Knowledge of Social Media," *The Information Society* 35, no. 1 (January 2019): 1–12, https://doi.org/10.1080/01972243.2018.1542647.

39. Acker and Donovan, "Data Craft"; Kevin Driscoll and Shawn Walker, "Big Data, Big Questions| Working Within a Black Box: Transparency in the Collection and Production of Big Twitter Data," *International Journal of Communication* 8 (June 16, 2014): 20; Shawn Walker, "The Complexity of Collecting Digital and Social Media Data in Ephemeral Contexts" (PhD thesis, University of Washington, 2017), https://digital.lib.washington.edu:443/researchworks/handle/1773/40612.

40. Using the API requires users to apply and register with a Facebook ID and agree to Facebook developers' terms of service. It can take up to two days for Facebook to verify a person's identity before giving them access to the API, and there is evidence that queries are monitored. At the time of writing, queries return up to five thousand ads, and unlike the global search feature, it is only available to search in markets in Brazil, Canada, EU, India, Israel, Ukraine, the UK, and the US. Facebook, "Facebook Ad Library API," 2020, https://www.facebook.com/ads/library/api.

41. Facebook, "Facebook Ad Library," 2020, https://www.facebook.com/ads/library/?active_status=all&ad_type=all&country=US&impression_search_field=has_impressions_lifetime.

42. Twitter Safety, "Disclosing New Data to Our Archive of Information Operations," Twitter (blog), September 20, 2019, https://blog.twitter.com/en_us/topics/company/2019/info-ops-disclosure-data-september-2019.html.

43. Twitter Inc., "Updating Our Advertising Policies on State Media," Twitter (blog), August 19, 2019, https://blog.twitter.com/en_us/topics/company/2019/advertising_policies_on_state_media.html.

44. Snap Inc., "Snap Political and Advocacy Ads Library—Business Center, Frequently Asked Questions," Snap Chat Business Center, 2020, https://businesshelp.snapchat.com/en-US/article/political-ads-library.

45. Snapchat's ad library is a notable exception in that it provides extensive metadata regarding age segmenting and advertising costs.

46. Researchers are also prohibited from depositing their data for preservation to ensure long-term access in a public data repository or data archive. However, some are beginning to explore federated and cooperative data archive models.

47. Mozilla, "Open Letter: Facebook, Do Your Part Against Disinformation," *The Mozilla Blog*, February 11, 2019, https://blog.mozilla.org/blog/2019/02/11/open-letter-facebook-do-your-part-against-disinformation.

48. Mozilla, "Facebook and Google: This Is What an Effective Ad Archive API Looks Like," *The Mozilla Blog*, March 27, 2019, https://blog.mozilla.org/blog/2019/03/27/facebook-and-google-this-is-what-an-effective-ad-archive-api-looks-like; Ashley Boyd, "Facebook Is Still Failing at Ad Transparency (No Matter What They Claim)," *The Mozilla Blog*, October 31, 2019, https://blog.mozilla.org/blog/2019/10/31/facebook-is-still-failing-at-ad-transparency-no-matter-what-they-claim.

49. Acker, "Data Craft"; Acker and Kriesberg, "Social Media Data Archives in an API-Driven World."

50. Paddy Leerssen et al., "Platform Ad Archives: Promises and Pitfalls," *Internet Policy Review* 8, no. 4 (October 9, 2019), https://policyreview.info/articles/analysis/platform-ad-archives-promises-and-pitfalls.

51. Vijaya Gadde and Yoel Roth, "Enabling Further Research of Information Operations on Twitter," Twitter (blog), October 17, 2018, https://blog.twitter.com/official/en_us/topics/company/2018/enabling-further-research-of-information-operations-on-twitter.html.

52. From 2018 to 2021, Twitter released datasets tied to manipulation campaigns and inauthentic behavior, many of which were state sponsored or found to be state linked. Twitter, "Disclosing State-Linked Information Operations We've Removed," Twitter (blog), December 2, 2021, https://blog.twitter.com/en_us/topics/company/2021/disclosing-state-linked-information-operations-we-ve-removed.

53. Gadde and Roth, "Enabling Further Research of Information Operations on Twitter."

54. Twitter Safety, "New Disclosures to Our Archive of State-Backed Information Operations," Twitter (blog), December 20, 2019, https://blog.twitter.com/en_us/topics/company/2019/new-disclosures-to-our-archive-of-state-backed-information-operations.html; Twitter Safety.

55. Marco Bastos, "This Account Doesn't Exist: Tweet Decay and the Politics of Deletion in the Brexit Debate," *American Behavioral Scientist* 65, no. 5 (May 1, 2021): 757–773, https://doi.org/10.1177/0002764221989772.

56. Daniel Carter et al., "Investigative Approaches to Researching Information Technology Companies," *Journal of the Association for Information Science and Technology* 72, no. 6 (June 2021): 655–666, https://doi.org/10.1002/asi.24446.

57. Siva Vaidhyanathan, *Antisocial Media: How Facebook Disconnects Us and Undermines Democracy* (New York: Oxford University Press, 2018); HKS Misinformation Review Guest Authors, "Tackling Misinformation: What Researchers Could Do with Social Media Data," *Harvard Kennedy School Misinformation Review*, December 9, 2020, https://doi.org/10.37016/mr-2020-49; Deen Freelon, "Chapter One When Social Media Data Disappear," in *Chapter One When Social Media Data Disappear* (New York: Columbia University Press, 2020), 6–29, https://doi.org/10.7312/harg18876-002.

58. Laura Edelson et al., "An Analysis of United States Online Political Advertising Transparency," *arXiv:1902.04385 [Cs]*, February 12, 2019, http://arxiv.org/abs/1902.04385; Mozilla, "Facebook and Google"; A. Rieke and M. Bogen, "Leveling the Platform: Real Transparency for Paid Messages on Facebook," *Upturn*, May 2018, https://www.upturn.org/work/leveling-the-platform.

59. Yelp, "Yelp Open Dataset: An All-Purpose Dataset for Learning," *Yelp*, 2020, https://www.yelp.com/dataset.

60. WhatsApp, "WhatsApp Research Awards for Social Science and Misinformation," WhatsApp.com, 2018, https://www.whatsapp.com/research/awards; Snap Inc., "2019 Snap Research Fellowship & Scholarship," 2019, https://snapresearchfs.splashthat.com.

61. Emily Taylor and Stacie Hoffman, "Industry Responses to Computational Propaganda and Social Media Manipulation," Working Paper (Oxford: Project on Computational Propaganda, 2019), https://comprop.oii.ox.ac.uk/research/industry-responses/.

62. Gary King and Nate Persily, "A New Model for Industry–Academic Partnerships," August 10, 2018, https://gking.harvard.edu/files/gking/files/partnerships.pdf.

63. Craig Silverman, "Facebook Said It Would Give Detailed Data to Academics. They're Still Waiting.," *BuzzFeed News*, August 23, 2019, https://www.buzzfeednews.com/article/craigsilverman/slow-facebook.

64. John and Nissenbaum, "An Agnotological Analysis of APIs."

65. There is some evidence that platforms are not able to accurately provide access to complete and accurate data, even under the direction of judges. For example, Facebook keeps user data managed in such a way that it cannot explain how it can be used. Lorenzo Franceschi-Bicchierai, "Facebook Doesn't Know What It Does with Your Data, or Where It Goes: Leaked Document," *Vice* (blog), April 26, 2022, https://www.vice.com/en/article/akvmke/facebook-doesnt-know-what-it-does-with-your-data-or-where-it-goes.

66. Social Science One, "Overview," *Social Science One*, 2018, https://socialscience.one/overview.

67. "Public Statement from the Co-Chairs and European Advisory Committee of Social Science One," December 11, 2019, https://socialscience.one/blog/public-statement-european-advisory-committee-social-science-one.

68. In an explanatory and critical perspective piece, de Vreese and Tromble write: "There is no guarantee that Facebook, nor any other company, will implement this model again, and it certainly does not solve the problem of inequitable access to data. Indeed, all researchers involved in the US 2020 project are from elite US universities." Claes de Vreese and Rebekah Tromble, "The Data Abyss: How Lack of Data Access Leaves Research and Society in the Dark," *Political Communication* 40, no. 3 (May 4, 2023): 356–360, https://doi.org/10.1080/10584609.2023.2207488.

69. House Committee on Science, Space, and Technology, "Investigations & Oversight Subcommittee Hearing—The Disinformation Black Box: Researching Social Media Data," House Committee on Science Space & Tech—Republicans, September 28, 2021, https://science.house.gov/2021/9/investigations-oversight-subcommittee-hearing-disinformation-black-box.

70. Davey Alba, "Facebook Sent Flawed Data to Misinformation Researchers," *New York Times*, February 24, 2022, sec. Technology, https://www.nytimes.com/live/2020/2020-election-misinformation-distortions.

71. Sarah Grevy Gotfredsen and Kaitlyn Dowling, "Meta Is Getting Rid of CrowdTangle—and Its Replacement Isn't as Transparent or Accessible," *Columbia Journalism Review*, July 9, 2024, https://www.cjr.org/tow_center/meta-is-getting-rid-of-crowdtangle.php.

72. Meghan Bobrowsky, "Facebook Disables Access for NYU Research into Political-Ad Targeting," *Wall Street Journal*, August 4, 2021, https://www.wsj.com/articles/facebook-cuts-off-access-for-nyu-research-into-political-ad-targeting-11628052204; Ariana Tobin and Jeremy B. Merrill, "Facebook Moves to Block Ad Transparency Tools—Including Ours," *ProPublica*, January 28, 2019, https://www.propublica.org/article/facebook-blocks-ad-transparency-tools.

73. Tromble, "Where Have All the Data Gone?"

74. Josh Constine, "Facebook and Instagram Launch US Political Ad Labeling and Archive," *TechCrunch* (blog), May 24, 2018, http://social.techcrunch.com/2018/05/24/facebook-political-ad-archive/; Vijaya Gadde and Bruce Falck, "Increasing Transparency for Political Campaigning Ads on Twitter," Twitter (blog), May 24, 2018, https://blog.twitter.com/en_us/topics/company/2018/Increasing-Transparency-for-Political-Campaigning-Ads-on-Twitter.html; Snap Inc., "Snap Political and Advocacy Ads Library—Business Center, Frequently Asked Questions."

75. Some social media researchers have encouraged pursuing less sanctioned methods such as brute force web scraping and violating platform's terms of service.

76. Andrew L. Russell, *Open Standards and the Digital Age: History, Ideology, and Networks* (New York: Cambridge University Press, 2014).

77. The White House, "FACT SHEET: United States and European Commission Announce Trans-Atlantic Data Privacy Framework," The White House, March 25, 2022, https://www.whitehouse.gov/briefing-room/statements-releases/2022/03/25/fact-sheet-united-states-and-european-commission-announce-trans-atlantic-data-privacy-framework/.

78. See, for example, the EU US Trans-Atlantic Data Privacy Framework that has been recently updated. Clothilde Goujard and Alfred Ng, "EU and US Reach a Deal to Let Data Flow across the Atlantic," *POLITICO* (blog), July 10, 2023, https://www.politico.eu/article/eu-signs-off-on-data-transfers-deal-with-us/.

79. Gilliland, "Reconceptualizing Records."

80. Acker, "A Death in the Timeline."

81. Lindsay Poirier, "Classification as Catachresis: Double Binds of Representing Difference with Semiotic Infrastructure," *Canadian Journal of Communication* 44, no. 3 (2019): 361–371; Lindsay Poirier, *Knowledge Representation in Scruffy Worlds: An Ethnography of Semiotic Infrastructure Design Work* (Troy, NY: Rensselaer Polytechnic Institute, 2018).

82. Roderic Crooks, "Representationalism at Work: Dashboards and Data Analytics in Urban Education," *Educational Media International* 54, no. 4 (October 2, 2017): 296–297, https://doi.org/10.1080/09523987.2017.1408267.

83. Andrew Iliadis, "The Tower of Babel Problem: Making Data Make Sense with Basic Formal Ontology," *Online Information Review* 43, no. 6 (January 1, 2019): 1021–1045, https://doi.org/10.1108/OIR-07-2018-0210.

84. Fidler and Acker, "Metadata, Infrastructure, and Computer-Mediated Communication in Historical Perspective."

85. Van Dijck, "Datafication, Dataism and Dataveillance."

86. Buckland has shown that metadata standards as infrastructure can be used to interpret patterns of thought and action among documents within the information systems that we use to create, organize, and name the human lifeworld. Buckland, "Description and Search."

87. Some metadata professionals have explored how user-generated metadata, such as dating profile terms and categories, can guide how we might theorize the use and potential flexibility of metadata for algorithmic sorting and matching in platforms. Deborah Maron and Erin Carter, "'More Than What It Seems': How Critical Theory, Popular Engagement and Apps Like Tinder Can Help Us Reframe Metadata and Its Consequences," in *Proceedings of the International Conference on Dublin Core and Metadata Applications*, Washington, DC, October 26–29, 2017.

88. MuckRock, "Palantir—DCAS," April 14, 2016, https://www.muckrock.com/foi/new-york-city-17/palantir-dcas-25162/.

89. Kelly Gates, "Policing as Digital Platform," *Surveillance & Society* 17, no. 1–2 (January 2019): 63–68, https://doi.org/10.24908/ss.v17i1/2.12940; Sarah Brayne, *Predict and Surveil: Data, Discretion, and the Future of Policing*, 1st ed. (New York: Oxford University Press, 2020).

90. William Alden, "There's A Fight Brewing Between The NYPD and Silicon Valley," *BuzzFeed News*, June 28, 2017, https://www.buzzfeednews.com/article/williamalden/theres-a-fight-brewing-between-the-nypd-and-silicon-valley.

91. Emily Hockett and Michael Price, "Palantir Contract Dispute Exposes NYPD's Lack of Transparency," *Just Security*, July 20, 2017, https://www.justsecurity.org/43397/palantir-contract-dispute-exposes-nypds-lack-transparency/.

92. It is worth noting that more popular, public definitions of metadata often appear via personal data breaches or surveillance overreach and tend to focus on ruptures and violations of individual users instead of the constant creation, tacit collection, and administrative cycles of metadata creation where data is being collected from networked groups, communities, and even populations (e.g., the Snowden leaks in 2013).

93. Anne Helmond, "The Platformization of the Web: Making Web Data Platform Ready," *Social Media + Society* 1, no. 2 (September 30, 2015), https://doi.org/10.1177/2056305115603080.

94. Gillespie, "The Politics of 'Platforms.'"

95. Arora, *The Next Billion Users*; Wendy Liu, *Abolish Silicon Valley: How to Liberate Technology from Capitalism* (London: Repeater, 2020).

96. Iain Munro, *Information Warfare in Business: Strategies of Control and Resistance in the Network Society* (London: Routledge, 2004), https://doi.org/10.4324/9780203449233; Gandy, *The Panoptic Sort*.

97. James R. Beniger, *The Control Revolution: Technological and Economic Origins of the Information Society* (Cambridge, MA: Harvard University Press, 1986); JoAnne Yates, *Control Through Communication: The Rise of System in American Management* (Baltimore, MD: Johns Hopkins University Press, 1993); Craig Robertson, "The Filing Cabinet," *Places Journal*, May 11, 2021, https://doi.org/10.22269/210511.

98. Thylstrup, *The Politics of Mass Digitization*, 5.

99. Tonia Sutherland, "Making a Killing: On Race, Ritual, and (Re)Membering in Digital Culture," *Preservation, Digital Technology & Culture* 46, no. 1 (2017): 32–40, https://doi.org/10.1515/pdtc-2017-0025.

100. In her work on information practices before and after disasters, Megan Finn has shown how private platforms such as Facebook convened groups of people in novel ways with algorithms, location data, and affinity groups. See Megan Finn,

Documenting Aftermath: Information Infrastructures in the Wake of Disasters (Cambridge, MA: MIT Press, 2018).

101. For more on platform accountability through public interest archives, please see Andrew Iliadis and Amelia Acker, "The Palantir Files: Public Interest Archives for Platform Accountability," *Information, Communication & Society* (June 2024): 1–23, https://doi.org/10.1080/1369118X.2024.2352624.

102. Leerssen et al., "Platform Ad Archives."

103. Philip M. Napoli, "User Data as Public Resource: Implications for Social Media Regulation," *Policy & Internet* 11, no. 4 (December 2019): 439–459, https://doi.org/10.1002/poi3.216.

104. Philip Mirowski, *Science-Mart: Privatizing American Science* (Cambridge, MA: Harvard University Press, 2011); Sarah Lamdan, *Data Cartels: The Companies That Control and Monopolize Our Information* (Stanford, CA: Stanford University Press, 2022).

105. Hon Vince Chhabria and Hon Jacqueline Scott Corley, Amended Order Regarding Production of Named Plaintiff Data, No. 3:18-MD-02843-VC-JSC (n.d.); Franceschi-Bicchierai, "Facebook Doesn't Know What It Does with Your Data, or Where It Goes."

106. Niels Brügger, "Web History and Social Media," *Sage Handbook of Social Media*, ed. Jean Burgess, Alice Marwick, and Thomas Poell (London: Sage, 2018), 196–212; Axel Bruns, "After the 'APIcalypse': Social Media Platforms and Their Fight against Critical Scholarly Research," *Information, Communication & Society* 22, no. 11 (September 19, 2019): 1544–1566, https://doi.org/10.1080/1369118X.2019.1637447; Robyn Caplan and danah boyd, "Isomorphism through Algorithms: Institutional Dependencies in the Case of Facebook," *Big Data & Society* 5, no. 1 (June 2018): 205395171875725, https://doi.org/10.1177/2053951718757253.

107. Acker and Flamm, "COVID-19 Community Archives and the Platformization of Digital Cultural Memory."

108. Randall C. Jimerson, "Embracing the Power of Archives," *The American Archivist* 69, no. 1 (2006): 19–32; Eric Ketelaar, "Archival Temples, Archival Prisons: Modes of Power and Protection," *Archival Science* 2, no. 3 (September 1, 2002): 221–238, https://doi.org/10.1007/BF02435623.

EPILOGUE

1. As many lawsuits claim, most of this data scraping is done without consent, attribution, citation, or notification of the people who created these works.

2. Agar, *Constant Touch*.

3. Some economists have suggested they have become "data monopolies."

4. For more on the data border milieu, see Melissa Villa-Nicholas, *Data Borders: How Silicon Valley is Building an Industry Around Immigrants* (Oakland, CA: University of California Press, 2023).

5. Brien Brothman, "The Limits of Limits: Derridean Deconstruction and the Archival Institution," *Archivaria* 36 (February 5, 1993): 205–220; Caswell, "'The Archive' Is Not an Archives."

6. Kean Birch et al., "Data as Asset? The Measurement, Governance, and Valuation of Digital Personal Data by Big Tech," *Big Data & Society* 8, no. 1 (January 1, 2021): 20539517211017308, https://doi.org/10.1177/20539517211017308.

7. Hogan and Roberts, "Archiving for Extinction."

BIBLIOGRAPHY

"550 DECTape Control Instruction Manual." Digital Equipment Corporation, 1966. Box 4A, Folder 5, 102652390. Computer History Museum, PDP-1 records.

Acker, Amelia. "A Death in the Timeline: Memory and Metadata in Social Platforms." *Journal of Critical Library and Information Studies* 2, no. 1 (2018): 27.

Acker, Amelia. "Accessing Software: Emulation in Information Institutions." *Information & Culture* 59, no. 1 (2023).

Acker, Amelia. "Data Craft: The Manipulation of Social Media Metadata." New York: Data & Society, November 5, 2018. https://datasociety.net/output/data-craft/.

Acker, Amelia. "Emulation Practices for Software Preservation in Libraries, Archives, and Museums." *Journal of the Association for Information Science and Technology* 72, no. 9 (2021): 1148–1160. https://doi.org/10.1002/asi.24482.

Acker, Amelia. "Metadata." In *Uncertain Archives*, edited by Nanna Bonde Thylstrup, Daniela Agostinho, Annie Ring, Catherine D'Ignazio, and Kristin Veel, 321–330. Cambridge, MA: MIT Press, 2021. https://doi.org/10.7551/mitpress/12236.003.0035.

Acker, Amelia. "The Short Message Service: Standards, Infrastructure and Innovation." *Telematics and Informatics* 31, no. 4 (November 1, 2014): 559–568. https://doi.org/10.1016/j.tele.2014.01.004.

Acker, Amelia, and Adam Kriesberg. "Social Media Data Archives in an API-Driven World." *Archival Science* 20 (2020): 105–123. https://doi.org/10.1007/s10502-019-09325-9.

Acker, Amelia, and Adam Kriesberg. "Tweets May Be Archived: Civic Engagement, Digital Preservation and Obama White House Social Media Data." *Proceedings of the Association for Information Science and Technology* 54, no. 1 (January 1, 2017): 1–9. https://doi.org/10.1002/pra2.2017.14505401001.

Acker, Amelia, and Joan Donovan. "Data Craft: A Theory/Methods Package for Critical Internet Studies." *Information, Communication & Society* 22, no. 11 (July 20, 2019): 1590–1609. https://doi.org/10.1080/1369118X.2019.1645194.

Acker, Amelia, and Lucy Flamm. "COVID-19 Community Archives and the Platformization of Digital Cultural Memory." In *Proceedings of the 54th Hawaii International Conference on System Sciences*, 2544–2553. Atlanta, GA: Association for Information Systems, 2021. http://hdl.handle.net/10125/70926.

Adavi, Krishna Akhil Kumar, and Amelia Acker. "What Is a File on a Phone? Personal Information Management Practices Amongst WhatsApp Users." *Proceedings of ACM Human–Computer Interaction* 7, no. 372 (2023): 1–28. https://doi.org/10.1145/3610221.

Agar, Jon. *Constant Touch: A Global History of the Mobile Phone*. London: Icon Books, 2013.

Agar, Jon. *The Government Machine: A Revolutionary History of the Computer*. Cambridge, MA: MIT Press, 2003.

Agre, Philip E. "Changing Places: Contexts of Awareness in Computing." *Human–Computer Interaction* 16 (2001): 177–192.

Akera, Atsushi. "Voluntarism and the Fruits of Collaboration: The IBM User Group, Share." *Technology and Culture* 42, no. 4 (2001): 710–736.

Alba, Davey. "Facebook Sent Flawed Data to Misinformation Researchers." *New York Times*, February 24, 2022, sec. Technology. https://www.nytimes.com/live/2020/2020-election-misinformation-distortions.

Alden, William. "There's A Fight Brewing Between The NYPD and Silicon Valley." *BuzzFeed News*, June 28, 2017. https://www.buzzfeednews.com/article/williamalden/theres-a-fight-brewing-between-the-nypd-and-silicon-valley.

Alpert-Abrams, Hannah, David A. Bliss, and Itza Carbajal. "Post-Custodial Archiving for the Collective Good: Examining Neoliberalism in US–Latin American Archival Partnerships." *Journal of Critical Library and Information Studies* 2, no. 1 (March 3, 2019), https://doi.org/10.24242/jclis.v2i1.87.

American Library Association. "Our Statement on Access." American Library Association, September 26, 2016. http://www.ala.org/rusa/strategic-priorities/access.

American Library Association. "Privacy: An Interpretation of the Library Bill of Rights." American Library Association, July 7, 2006. http://www.ala.org/advocacy/intfreedom/librarybill/interpretations/privacy.

Andrews, Dave. "Personal Digital Assistants: Behind the Wheel of the First Zoomer and Newton PDAs." *BYTE*, September 1, 1993.

Anonymous. "Portable and Personal: Take Your Office with You." *Managing Office Technology* 38, no. 11 (November 1993): 49.

Arora, Payal. *The Next Billion Users: Digital Life Beyond the West*. Cambridge, MA: Harvard University Press, 2019.

Atkin, Danny. "Casio Z-7000; Tandy Z-PDA. (Personal Digital Assistants) Hardware Review." *COMPUTE!*, March 1994.

Au, Yung. "Data Centres on the Moon and Other Tales: A Volumetric and Elemental Analysis of the Coloniality of Digital Infrastructures." *Territory, Politics, Governance* 12, no. 1 (2024): 12–30.

Ayto, John. *Dictionary of Word Origins*. New York: Arcade, 2011.

Banoub, Daniel, and Sarah J. Martin. "Storing Value: The Infrastructural Ecologies of Commodity Storage." *Environment and Planning D: Society and Space* 38, no. 6 (December 1, 2020): 1101–1019. https://doi.org/10.1177/0263775820911942.

Barnhart, Robert K. *Chambers Dictionary of Etymology*. Edinburgh, UK: Chambers, 1999.

Bastos, Marco. "This Account Doesn't Exist: Tweet Decay and the Politics of Deletion in the Brexit Debate." *American Behavioral Scientist* 65, no. 5 (May 1, 2021): 757–773. https://doi.org/10.1177/0002764221989772.

Beck, Carl. "In Memoriam: Ralph L. Bisco." *PS: Political Science & Politics* 3, no. 3 (1970): 467–467. https://doi.org/10.1017/S1049096500029413.

Bell, C. Gordon, and Michael M. Gold. "An Introduction to the Structure of Time-Shared Computers." In *Advances in Information Systems Science: Volume 4*, edited by Julius T. Tou, 161–272. Boston, MA: Springer US, 1972. https://doi.org/10.1007/978-1-4615-9053-8_4.

Bell, C. Gordon, J. Craig Mudge, and John E. McNamara. *Computer Engineering: A DEC View of Hardware Systems Design*. Bedford, MA: Digital Press, 2014.

Bell, Gordon. "Computer Generations." Digital Equipment Corporation, 1975. Box 60, Folder 4. Computer History Museum, Digital Equipment Corporation records.

Bell, Gordon. "Toward a History of (Personal) Workstations." In *Proceedings of the ACM Conference on The History of Personal Workstations*, 1–17. HPW '86. New York: Association for Computing Machinery, 1986. https://doi.org/10.1145/12178.12179.

Bengtsson, Linus, Xin Lu, Anna Thorson, Richard Garfield, and Johan von Schreeb. "Improved Response to Disasters and Outbreaks by Tracking Population Movements with Mobile Phone Network Data: A Post-Earthquake Geospatial Study in Haiti." *PLOS Medicine* 8, no. 8 (August 30, 2011): e1001083. https://doi.org/10.1371/journal.pmed.1001083.

Beniger, James R. *The Control Revolution: Technological and Economic Origins of the Information Society*. Cambridge, MA: Harvard University Press, 1986.

Benthall, Sebastian, Seda Gürses, and Helen Nissenbaum. "Contextual Integrity Through the Lens of Computer Science." *Foundations and Trends in Privacy and Security* 2, no. 1 (December 22, 2017): 1–69. https://doi.org/10.1561/3300000016.

Bergman, Ofer, and Steve Whittaker. *The Science of Managing Our Digital Stuff Ofer Bergman and Steve Whittaker*. Cambridge, MA: MIT Press, 2016.

Birch, Kean. "Technoscience Rent: Toward a Theory of Rentiership for Technoscientific Capitalism." *Science, Technology, & Human Values* 45, no. 1 (January 1, 2020): 3–33. https://doi.org/10.1177/0162243919829567.

Birch, Kean, D. T. Cochrane, and Callum Ward. "Data as Asset? The Measurement, Governance, and Valuation of Digital Personal Data by Big Tech." *Big Data & Society* 8, no. 1 (January 1, 2021): 20539517211017308. https://doi.org/10.1177/20539517211017308.

Bisco, Ralph L. "Social Science Data Archives: A Review of Developments." *The American Political Science Review* 60, no. 1 (1966): 93–109. https://doi.org/10.2307/1953810.

Bisco, Ralph L. "Social Science Data Archives: Progress and Prospects." *Social Science Information* 6, no. 1 (February 1, 1967): 39–74. https://doi.org/10.1177/053901846700600102.

Bisco, Ralph L. "Social Science Data Archives Technical Considerations." *Social Science Information* 4, no. 3 (September 1, 1965): 129–150. https://doi.org/10.1177/053901846500400311.

Bisco, Ralph L., and Council of Social Science Data Archives. *Data Bases, Computers, and the Social Sciences*. Edited by Ralph L. Bisco. Information Sciences Series. New York: Wiley-Interscience, 1970.

Blondel, Vincent D., Adeline Decuyper, and Gautier Krings. "A Survey of Results on Mobile Phone Datasets Analysis." *EPJ Data Science* 4, no. 1 (December 2015): 1–55. https://doi.org/10.1140/epjds/s13688-015-0046-0.

Blumenthal, Marcia. "Fraud Ruled in 1968 DEC Tape Patent." *Computerworld*, May 1, 1978.

Bobrowsky, Meghan. "Facebook Disables Access for NYU Research into Political-Ad Targeting." *Wall Street Journal*, August 4, 2021. https://www.wsj.com/articles/facebook-cuts-off-access-for-nyu-research-into-political-ad-targeting-11628052204.

Borgman, Christine L. *Big Data, Little Data, No Data: Scholarship in the Networked World*. Cambridge, MA: MIT Press, 2015.

Borgman, Christine L. *Scholarship in the Digital Age: Information, Infrastructure, and the Internet*. Cambridge, MA: MIT Press, 2007.

Bouk, Dan. "The National Data Center and the Rise of the Data Double." *Historical Studies in the Natural Sciences* 48, no. 5 (November 1, 2018): 627–636. https://doi.org/10.1525/hsns.2018.48.5.627.

Bowman, Raymond T. "Crossroad Choices for the Future Development of the Federal Statistical System." *Journal of the American Statistical Association* 63, no. 323 (1968): 801–816. https://doi.org/10.2307/2283874.

Bowman, Raymond T. "The Idea of a Federal Statistical Data Center—It's Purposes and Structure." In *Data Bases, Computers, and the Social Sciences*, edited by Ralph L. Bisco, 63–69. Information Sciences Series. New York: Wiley-Interscience, 1970.

Boyd, Ashley. "Facebook Is Still Failing at Ad Transparency (No Matter What They Claim)." *The Mozilla Blog*, October 31, 2019. https://blog.mozilla.org/blog/2019/10/31/facebook-is-still-failing-at-ad-transparency-no-matter-what-they-claim.

Braman, Sandra. *Change of State: Information, Policy, and Power*. Cambridge, MA: MIT Press, 2009.

Brayne, Sarah. *Predict and Surveil: Data, Discretion, and the Future of Policing*. 1st ed. New York: Oxford University Press, 2020.

Brothman, Brien. "The Limits of Limits: Derridean Deconstruction and the Archival Institution." *Archivaria* 36(February 5, 1993): 205–220.

Brügger, Niels. "Web History and Social Media." In *Sage Handbook of Social Media*, edited by Jean Burgess, Alice Marwick, and Thomas Poell, 196–212. London: Sage, 2018.

Bruns, Axel. "After the 'APIcalypse': Social Media Platforms and Their Fight against Critical Scholarly Research." *Information, Communication & Society* 22, no. 11 (September 19, 2019): 1544–1566. https://doi.org/10.1080/1369118X.2019.1637447.

Buckland, Michael K. "Description and Search: Metadata as Infrastructure." *Brazilian Journal of Information Science: Research Trends* 0, no. 0 (March 20, 2007). https://doi.org/10.36311/1981-1640.2006.V0N0.02.P3.

Buckland, Michael K. "Information as Thing." *Journal of the American Society for Information Science* 42, no. 5 (June 1, 1991): 351–360. https://doi.org/10.1002/(SICI)1097-4571(199106)42:5<351:AID-ASI5>3.0.CO;2-3.

Butter, Andrea, and David Pogue. *Piloting Palm: The Inside Story of Palm, Handspring, and the Birth of the Billion-Dollar Handheld Industry*. 1st ed. New York, NY: Wiley, 2002.

Campbell-Kelly, Martin, William Aspray, Nathan Ensmenger, and Jeffrey R. Yost. *Computer: A History of the Information Machine*. 3rd ed. New York: Routledge, 2019. https://doi.org/10.4324/9780429495373.

Caplan, Robyn, and danah boyd. "Isomorphism through Algorithms: Institutional Dependencies in the Case of Facebook." *Big Data & Society* 5, no. 1 (June 2018): 205395171875725. https://doi.org/10.1177/2053951718757253.

Carter, Daniel, Amelia Acker, and Dan Sholler. "Investigative Approaches to Researching Information Technology Companies." *Journal of the Association for Information Science and Technology* 72, no. 6 (June 2021): 655–666. https://doi.org/10.1002/asi.24446.

Carter, Luther J. "National Data Bank: Its Advocates Try to Erase 'Big Brother' Image." *Science* 163, no. 3863 (1969): 160–162.

Caswell, M. L. "'The Archive' Is Not an Archives: On Acknowledging the Intellectual Contributions of Archival Studies." June 27, 2021. https://escholarship.org/uc/item/7bn4v1fk.

Ceruzzi, Paul E. *A History of Modern Computing*. Edited by William Aspray. 2nd ed. Cambridge, MA: MIT Press, 2003.

Cetina, Karin Knorr. *Epistemic Cultures: How the Sciences Make Knowledge*. Cambridge, MA: Harvard University Press, 1999.

Chambers, Sally, Melanie Walsh, Michelle Caswell, et al. "Position Statements -> Collections as Data: State of the Field and Future Directions." May 2, 2023. https://doi.org/10.5281/zenodo.7897735.

Chappell, Bill. "Obama: Cyberspace Is the New 'Wild West.'" *NPR*, February 13, 2015. https://www.npr.org/sections/thetwo-way/2015/02/13/385960693/obama-to-urge-companies-to-share-data-on-cyber-threats.

Chhabria, Hon Vince, and Hon Jacqueline Scott Corley. Amended Order Regarding Production of Named Plaintiff Data No. 3:18-MD-02843-VC-JSC. (n.d.).

Christopher, A. "Wherefore Art Thou, Apple?" *Upside*, October 1991.

Chun, Wendy Hui Kyong. *Control and Freedom: Power and Paranoia in the Age of Fiber Optics*. Cambridge, MA: MIT Press, 2006.

Chun, Wendy Hui Kyong. *Programmed Visions: Software and Memory*. Cambridge, MA: MIT Press, 2011.

Chun, Wendy Hui Kyong. "The Enduring Ephemeral, or the Future Is a Memory." *Critical Inquiry* 35, no. 1 (2008): 148–171. https://doi.org/10.1086/595632.

Chun, Wendy Hui Kyong. *Updating to Remain the Same: Habitual New Media*. Cambridge, MA: MIT Press, 2016.

Ciborra, Claudio. *The Labyrinths of Information: Challenging the Wisdom of Systems*. 1st ed. New York: Oxford University Press, 2004.

Columbia University Bureau of Applied Social Research. *Specialized Social Science Information Services in the United States/Jack Ferguson*. New York: Distributed by Clearinghouse for Federal Scientific and Technical Information, 1965.

Computer Chronicles 1993 PDA Review Part 3 of 5, 2009. https://www.youtube.com/watch?v=BPt-SOtnyE0.

Constine, Josh. "Facebook and Instagram Launch US Political Ad Labeling and Archive." *TechCrunch* (blog), May 24, 2018. http://social.techcrunch.com/2018/05/24/facebook-political-ad-archive/.

Cook, Terry. "'We Are What We Keep; We Keep What We Are': Archival Appraisal Past, Present and Future." *Journal of the Society of Archivists* 32, no. 2 (October 2011): 173–189. https://doi.org/10.1080/00379816.2011.619688.

Corbató, Fernando, Marjorie Merwin-Dagget, and Robert Daley. "An Experimental Time-Sharing System." *Proceedings of the AIEE-IRE*, May 1, 1962. https://dl.acm.org/doi/abs/10.1145/1460833.1460871.

Cortada, James W. *Before the Computer: IBM, NCR, Burroughs, & Remington Rand & the Industry They Created, 1865–1956*. Princeton, NJ: Princeton University Press, 2000.

Cortada, James W. *Historical Dictionary of Data Processing: Technology*. New York: Greenwood, 1987.

Cortes, Corinna, Daryl Pregibon, and Chris Volinsky. "Computational Methods for Dynamic Graphs." *Journal of Computational and Graphical Statistics* 12, no. 4 (December 1, 2003): 950–970. https://doi.org/10.1198/1061860032742.

Crooks, Roderic. "Representationalism at Work: Dashboards and Data Analytics in Urban Education." *Educational Media International* 54, no. 4 (October 2, 2017): 289–303. https://doi.org/10.1080/09523987.2017.1408267.

Day, Ronald E. *Indexing It All: The Subject in the Age of Documentation, Information, and Data*. 1st ed. Cambridge, MA: MIT Press, 2014.

Day, Ronald E. *The Modern Invention of Information*. Carbondale: South Illinois University Press, 2008.

"DECtape Control 551 Preliminary Manual." Digital Equipment Corporation, 1965. Box 179, Folder 14. Computer History Museum, Digital Equipment Corporation records. http://www.bitsavers.org/pdf/dec/dectape/551/H-551P_DECtape_Control_551_Preliminary.pdf.

Dennis, Jack B. "A Time-Shared Computation Facility for Education and Research." Massachusetts Institute of Technology, May 16, 1963. Box 3, Folder 8. Computer History Museum, PDP-1 records.

Desrosières, Alain. *The Politics of Large Numbers: A History of Statistical Reasoning*. Translated by Camille Naish. Cambridge, MA: Harvard University Press, 2010.

de Vreese, Claes, and Rebekah Tromble. "The Data Abyss: How Lack of Data Access Leaves Research and Society in the Dark." *Political Communication* 40, no. 3 (May 4, 2023): 356–360. https://doi.org/10.1080/10584609.2023.2207488.

Diaz, Jaclyn. "Special Counsel Jack Smith Got a Secret Search Warrant for Trump's Twitter Account." *NPR*, August 9, 2023, sec. Law. https://www.npr.org/2023/08/09/1192953166/prosecutors-search-warrant-trump-twitter-account.

Digital Equipment Corporation. *PDP11: TC11 DECtape System Manual*. Maynard, MA: Digital Equipment Corporation, 1971.

Digital Public Library of America. "Philosophy—Why Build the DPLA API?" Digital Public Library of America, 2019. https://pro.dp.la/developers/philosophy.

Dollar, Charles M. "Computers, the National Archives, and Researchers." *Prologue-Quarterly of the National Archives* 8, no. 1 (1976): 29–34.

Dourish, Paul. *The Stuff of Bits: An Essay on the Materialities of Information*. Cambridge, MA: MIT Press, 2017.

Driscoll, Kevin. "From Punched Cards to 'Big Data': A Social History of Database Populism." *Communication 1* 1, no. 1 (August 29, 2012).

Driscoll, Kevin. *The Modem World: A Prehistory of Social Media*. 1st ed. New Haven, CT: Yale University Press, 2022.

Driscoll, Kevin, and Shawn Walker. "Big Data, Big Questions| Working Within a Black Box: Transparency in the Collection and Production of Big Twitter Data." *International Journal of Communication* 8 (June 16, 2014): 20.

Dunn, Edgar S. "Review of Proposal for a National Data Center, Statistical Evaluation Report." US Bureau of the Budget, Office of Statistical Standards, December 1965.

Dunn, Edgar S. "The Idea of a National Data Center and the Issue of Personal Privacy." *The American Statistician* 21, no. 1 (1967): 21–27. https://doi.org/10.2307/2681910.

Edelson, Laura, Shikhar Sakhuja, Ratan Dey, and Damon McCoy. "An Analysis of United States Online Political Advertising Transparency." *arXiv:1902.04385 [Cs]*, February 12, 2019. http://arxiv.org/abs/1902.04385.

Edwards, Paul N. *The Closed World: Computers and the Politics of Discourse in Cold War America*. Reprint ed. Cambridge, MA: MIT Press, 1997.

Electromechanical Design: Components & Systems. "DECtape Transport Design." November 1964. Box 179, Folder 12. Computer History Museum, Digital Equipment Corporation records. https://deramp.com/downloads/mfe_archive/011-Digital%20Equipment%20Corporation/20%20Magnetic%20Tape%20Subsystems/dectapeDesignNov64.pdf.

Eng, Paul, and Amy Dunkin. "Smart, Useful—and They Won't Put a Sag in Your Suit." *BusinessWeek*, no. 3374 (May 30, 1994): 141–142.

Ernst, Wolfgang. *Stirrings in the Archives: Order from Disorder*. Translated by Adam Siegel. Lanham, MD: Rowman & Littlefield, 2015.

Executive Office of the President, Office of Science and Technology. "Privacy and Behavioral Research." Washington, DC, February 1967.

Facebook. "Facebook Ad Library." Facebook, 2020. https://www.facebook.com/ads/library/?active_status=all&ad_type=all&country=US&impression_search_field=has_impressions_lifetime.

Facebook. "Facebook Ad Library API." Facebook, 2020. https://www.facebook.com/ads/library/api.

"Federal Data Banks, Computers and the Bill of Rights. Hearings before the Subcommittee on Constitutional Rights of the Committee on the Judiciary, Part 1." Washington, DC, 1971.

Feinberg, Melanie. "Reading Databases: Slow Information Interactions beyond the Retrieval Paradigm." *Journal of Documentation* 73, no. 2 (February 22, 2017): 336–356. https://doi.org/10.1108/JD-03-2016-0030.

Feinberg, Melanie. Melanie. *Everyday Adventures with Unruly Data*. Cambridge, MA: MIT Press, 2022.

Ferriero, David S. "Letter to Maloney on Trump Presidential Records on Social Media Platforms." February 18, 2022. https://www.archives.gov/files/foia/ferriero-letter-to-maloney-on-trump-presidential-records-on-social-media-platforms.02.18.2022.pdf.

Fidler, Bradley, and Amelia Acker. "Metadata, Infrastructure, and Computer-Mediated Communication in Historical Perspective." *Journal of the Association for Information Science and Technology* 68, no. 2 (February 1, 2017): 412–422. https://doi.org/10.1002/asi.23660.

Finn, Megan. *Documenting Aftermath: Information Infrastructures in the Wake of Disasters*. Cambridge, MA: MIT Press, 2018.

Fishbein, Meyer H. "Automation in Archives: A Summary History." National Archives. Maryland, 1981.

Fondren, Elisabeth, and McCune Meghan Menard. "Archiving and Preserving Social Media at the Library of Congress: Institutional and Cultural Challenges to Build a Twitter Archive." *Preservation, Digital Technology & Culture* 47, no. 2 (2018): 33–44. https://doi.org/10.1515/pdtc-2018-0011.

Franceschi-Bicchierai, Lorenzo. "Facebook Doesn't Know What It Does with Your Data, or Where It Goes: Leaked Document." *Vice* (blog), April 26, 2022. https://www.vice.com/en/article/akvmke/facebook-doesnt-know-what-it-does-with-your-data-or-where-it-goes.

Fredkin, Edward. "The Time Sharing of Computers." *Computers and Automation*, November 1963.

Freelon, Deen. "Chapter One When Social Media Data Disappear." In *Chapter One When Social Media Data Disappear*, 6–29. New York: Columbia University Press, 2020. https://doi.org/10.7312/harg18876-002.

Freelon, Deen. "Computational Research in the Post-API Age." *Political Communication* 35, no. 4 (October 2, 2018): 665–668. https://doi.org/10.1080/10584609.2018.1477506.

Gadde, Vijaya, and Bruce Falck. "Increasing Transparency for Political Campaigning Ads on Twitter." Twitter (blog), May 24, 2018. https://blog.twitter.com/en_us/topics/company/2018/Increasing-Transparency-for-Political-Campaigning-Ads-on-Twitter.html.

Gadde, Vijaya, and Yoel Roth. "Enabling Further Research of Information Operations on Twitter." Twitter (blog), October 17, 2018. https://blog.twitter.com/official/en_us/topics/company/2018/enabling-further-research-of-information-operations-on-twitter.html.

Gallagher, Cornelius. "Privacy and the National Data Center." Speech presented at the Joint Computer Conference, Atlantic City, NJ, April 18, 1967.

Gandy, Oscar H., Jr. *The Panoptic Sort: A Political Economy of Personal Information*. Boulder, CO: Westview Press, 1993.

Gates, Kelly. "Policing as Digital Platform." *Surveillance & Society* 17, no. 1–2 (January 2019): 63–68. https://doi.org/10.24908/ss.v17i1/2.12940.

Geda, Carolyn L. "Social Science Data Archives." *The American Archivist* 42, no. 2 (1979): 158–166.

Giannachi, Gabriella. *Archive Everything: Mapping the Everyday*, 1st ed. Cambridge, MA: MIT Press, 2016.

Gillespie, Tarleton. "The Politics of 'Platforms.'" *New Media & Society* 12, no. 3 (May 1, 2010): 347–364. https://doi.org/10.1177/1461444809342738.

Gilliland, Anne J. *Conceptualizing 21st-Century Archives*. Chicago, IL: Society of American Archivists, 2017.

Gilliland, Anne J. "Reconceptualizing Records, the Archive and Archival Roles and Requirements in a Networked Society." *Book Science* 63, no. 63 (January 1, 2014): 17–34. https://doi.org/10.15388/Knygotyra.2014.63.4011.

Gilliland, Anne J. "Reflections on the Value of Metadata Archaeology for Recordkeeping in a Global, Digital World." *Journal of the Society of Archivists* 32, no. 1 (April 1, 2011): 103–118. https://doi.org/10.1080/00379816.2011.563934.

Gilliland, Anne J. "Setting the Stage." In *Introduction to Metadata*, edited by Murtha Baca, 3rd ed., 1–19. Los Angeles, CA: J. Paul Getty Trust, 2016. http://www.getty.edu/publications/intrometadata.

Glazer, Rashi, B. J. Asirvatham, Tim Mai, and Narayana Mani. *Palm Computing in 1994*. London: Sage, 2000. https://doi.org/10.4135/9781526410788.

Gotfredsen, Sarah Grevy, and Kaitlyn Dowling. "Meta Is Getting Rid of CrowdTangle—and Its Replacement Isn't as Transparent or Accessible." *Columbia Journalism Review*, July 9, 2024. https://www.cjr.org/tow_center/meta-is-getting-rid-of-crowdtangle.php.

Goujard, Clothilde, and Alfred Ng. "EU and US Reach a Deal to Let Data Flow across the Atlantic." *POLITICO* (blog), July 10, 2023. https://www.politico.eu/article/eu-signs-off-on-data-transfers-deal-with-us/.

Greene, Daniel. *The Promise of Access*. Cambridge, MA: MIT Press, 2021. https://mitpress.mit.edu/9780262542333/the-promise-of-access/.

Greenwald, Glenn. "NSA Collecting Phone Records of Millions of Verizon Customers Daily." *The Guardian*, June 6, 2013, sec. US news. https://www.theguardian.com/world/2013/jun/06/nsa-phone-records-verizon-court-order.

Guay, Rob, and Kean Birch. "A Comparative Analysis of Data Governance: Socio-Technical Imaginaries of Digital Personal Data in the USA and EU (2008–2016)." *Big Data & Society* 9, no. 2 (July 1, 2022): 20539517221112925. https://doi.org/10.1177/20539517221112925.

Haigh, Thomas. "How Data Got Its Base: Information Storage Software in the 1950s and 1960s." *IEEE Annals of the History of Computing* 31, no. 4 (October 2009): 6–25. https://doi.org/10.1109/MAHC.2009.123.

Harper, Richard, Siân Lindley, Eno Thereska, et al. "What Is a File?" In *Proceedings of the 2013 Conference on Computer Supported Cooperative Work*, 1125–1136. CSCW '13. New York: Association for Computing Machinery, 2013. https://doi.org/10.1145/2441776.2441903.

Harrison, Joseph Leroy. "The Public Library Movement in the United States." *The Library* 1–8, no. 1 (January 1, 1896): 141–149. https://doi.org/10.1093/library/s1-8.1.141.

Helmond, Anne. "The Platformization of the Web: Making Web Data Platform Ready." *Social Media + Society* 1, no. 2 (September 30, 2015). https://doi.org/10.1177/2056305115603080.

Hill, G. Christian, and Jim Carlton. "Telecommunications (A Special Report): A Wireless World—Getting Personal: The Newton Has Given Personal Digital Assistants a Bad Name; But Don't Be Fooled: PDAs Will Soon Revolutionize the Way People Communicate." *Wall Street Journal*, Eastern Edition, February 11, 1994.

Hinz, H. C. "Magnetic Tape Technology in the 1990s." In *[1990] Digest of Papers. Tenth IEEE Symposium on Mass Storage Systems@m_Crisis in Mass Storage*, Monterey, CA, 38–42. New York: IEEE, 1990. https://doi.org/10.1109/MASS.1990.113566.

HKS Misinformation Review Guest Authors. "Tackling Misinformation: What Researchers Could Do with Social Media Data." *Harvard Kennedy School Misinformation Review*, December 9, 2020. https://doi.org/10.37016/mr-2020-49.

Hobart, Michael E., and Zachary S. Schiffman. *Information Ages: Literacy, Numeracy, and the Computer Revolution*. Baltimore: Johns Hopkins University Press, 2000.

Hockett, Emily, and Michael Price. "Palantir Contract Dispute Exposes NYPD's Lack of Transparency." *Just Security*, July 20, 2017. https://www.justsecurity.org/43397/palantir-contract-dispute-exposes-nypds-lack-transparency/.

Hogan, Mél. "Facebook Data Storage Centers as the Archive's Underbelly." *Television & New Media* 16, no. 1 (January 1, 2015): 3–18. https://doi.org/10.1177/1527476413509415.

Hogan, Mél, and Sarah T. Roberts. "Archiving for Extinction." *Media-N* 19, no. 1 (February 3, 2023): 7–26. https://doi.org/10.21900/j.median.v19i1.936.

Hopkins, Judith. "The 1791 French Cataloging Code and the Origins of the Card Catalog." *Libraries & Culture* 27, no. 4 (1992): 378–404.

Horrigan, John. "Consumption of Information Goods and Services in the U.S." *Pew Research Center: Internet, Science & Tech* (blog), November 23, 2003. https://www.pewresearch.org/internet/2003/11/23/consumption-of-information-goods-and-services-in-the-u-s/.

House Committee on Science, Space, and Technology. "Investigations & Oversight Subcommittee Hearing—The Disinformation Black Box: Researching Social Media Data." House Committee on Science Space & Tech—Republicans, September 28, 2021. https://science.house.gov/2021/9/investigations-oversight-subcommittee-hearing-disinformation-black-box.

Hu, Tung-Hui. *A Prehistory of the Cloud*. Cambridge, MA: MIT Press, 2015.

Igo, Sarah E. *The Known Citizen: A History of Privacy in Modern America*. Illustrated ed. Cambridge, MA: Harvard University Press, 2018.

Iliadis, Andrew. "The Tower of Babel Problem: Making Data Make Sense with Basic Formal Ontology." *Online Information Review* 43, no. 6 (January 1, 2019): 1021–1045. https://doi.org/10.1108/OIR-07-2018-0210.

Iliadis, Andrew, and Amelia Acker. "The Palantir Files: Public Interest Archives for Platform Accountability." *Information, Communication & Society* (June 2024): 1–23. https://doi.org/10.1080/1369118X.2024.2352624.

Iron Mountain. "Tape Backup Advantages for Ransomware Attacks." *Iron Mountain*, August 2021. https://www.ironmountain.com/resources/general-articles/t/tape-backup-advantages-for-ransomware-attacks.

Jackson, Steven J., Paul N. Edwards, Geoffrey C. Bowker, and Cory P. Knobel. "Understanding Infrastructure: History, Heuristics, and Cyberinfrastructure Policy." *First Monday* 12, no. 6 (2007): 1–9.

Jerney, J. "Talking with Palm Chair Jeff Hawkins." *Pen Computing*, January 1995.

Jimerson, Randall C. "Embracing the Power of Archives." *The American Archivist* 69, no. 1 (2006): 19–32.

John, Nicholas A., and Asaf Nissenbaum. "An Agnotological Analysis of APIs: Or, Disconnectivity and the Ideological Limits of Our Knowledge of Social Media." *The Information Society* 35, no. 1 (January 2019): 1–12. https://doi.org/10.1080/01972243.2018.1542647.

Julyk, David P., "The Trouble with Machines Is People: The Computer as Icon in Post-War America: 1946-1970." PhD dissertation, The University of Michigan, 2008.

Karger, David E. "Unify Everything: It's All the Same to Me." In *Personal Information Management*, edited by William P. Jones and Jaime Teevan, 127–152. Seattle: University of Washington Press, 2007. http://site.ebrary.com/id/10448623.

Kaysen, Carl, Charles C. Holt, Richard Holton, George Kozmetsky, H. Russell Morrison, and Richard Ruggles. "Report of the Task Force on the Storage of and Access to Government Statistics." *The American Statistician* 23, no. 3 (1969): 11–19.

Kelly, Thomas. *Early Public Libraries: A History of Public Libraries in Great Britain Before 1850*. London: Library Association, 1966.

Kelty, Christopher M. *Two Bits: The Cultural Significance of Free Software*. Experimental Futures. Durham, NC: Duke University Press, 2008.

Ketelaar, Eric. "Archival Temples, Archival Prisons: Modes of Power and Protection." *Archival Science* 2, no. 3 (September 1, 2002): 221–238. https://doi.org/10.1007/BF02435623.

King, Gary, and Nate Persily. "A New Model for Industry–Academic Partnerships," August 10, 2018. https://gking.harvard.edu/files/gking/files/partnerships.pdf.

Kirschenbaum, Matthew. *Mechanisms: New Media and the Forensic Imagination*. Cambridge, MA: MIT Press, 2008.

Kitchin, Rob. *The Data Revolution: Big Data, Open Data, Data Infrastructures and Their Consequences*. 1st ed. Los Angeles, CA: Sage, 2014.

Klein, Adam I., Jane E. Nitze, Edward W. Felten, Travis LeBlanc, and Aditya Bamzai. "Privacy and Civil Liberties Oversight Board Report on the Government's Use of the Call Detail Records Program Under the USA Freedom Act." February 2020, 106.

Klonick, Kate. "The New Governors: The People, Rules, and Processes Governing Online Speech." *Harvard Law Review* 131, no. 6 (2018): 1598–1670.

Kraus, Rebecca. "Statistical Déjà Vu: The National Data Center Proposal of 1965 and Its Descendants." *Journal of Privacy and Confidentiality* 5, no. 1 (August 1, 2013). https://doi.org/10.29012/jpc.v5i1.624.

Kriesberg, Adam, and Amelia Acker. "The Second US Presidential Social Media Transition: How Private Platforms Impact the Digital Preservation of Public Records." *Journal of the Association for Information Science and Technology* 73, no. 11 (2022): 1529–1542. https://doi.org/10.1002/asi.24659.

Kriesberg, Adam, Kerry Huller, Ricardo Punzalan, and Cynthia Parr. "An Analysis of Federal Policy on Public Access to Scientific Research Data." *Data Science Journal* 16 (June 14, 2017): 27. https://doi.org/10.5334/dsj-2017-027.

Kris, David S. "The NSA and the USA Freedom Act." *Lawfare* (blog), July 2, 2018. https://www.lawfareblog.com/nsa-and-usa-freedom-act.

Lamdan, Sarah. *Data Cartels: The Companies That Control and Monopolize Our Information*. Stanford, CA: Stanford University Press, 2022.

Landau, Susan. "Transactional Information Is Remarkably Revelatory." *Proceedings of the National Academy of Sciences* 113, no. 20 (2016): 5467–5469.

Landau, Susan, and Asaf Lubin. "Examining the Anomalies, Explaining the Value: Should the USA FREEDOM Act's Metadata Program Be Extended?" *Harvard National Security Journal* 11 (2020): 308–358.

Lauer, Josh. *Creditworthy: A History of Consumer Surveillance and Financial Identity in America*. New York: Columbia University Press, 2017.

Leerssen, Paddy, Jef Ausloos, Brahim Zarouali, Natali Helberger, and Claes H. de Vreese. "Platform Ad Archives: Promises and Pitfalls." *Internet Policy Review* 8, no. 4 (October 9, 2019). https://policyreview.info/articles/analysis/platform-ad-archives-promises-and-pitfalls.

Lehuedé, Sebastián. "Territories of Data: Ontological Divergences in the Growth of Data Infrastructure." *Tapuya: Latin American Science, Technology and Society* 5, no. 1 (2022). https://doi.org/10.1080/25729861.2022.2035936.

Lenstra, Noah. "The Library as Playground: How Games and Play Are Reshaping Public Culture." *Journal of Education for Library and Information Science* 64, no. 1 (January 2023): 102–103. https://doi.org/10.3138/jelis.2022-0041.

Leonelli, Sabina. "What Counts as Scientific Data? A Relational Framework." *Philosophy of Science* 82, no. 5 (December 2015): 810–821. https://doi.org/10.1086/684083.

Lepore, Jill. *If Then: How the Simulmatics Corporation Invented the Future*. New York: Liveright, 2020.

Levine, Yasha. *Surveillance Valley: The Secret Military History of the Internet*. Illustrated ed. New York: PublicAffairs, 2018.

Levy, Steven. *Hackers: Heroes of the Computer Revolution*, 1st ed. Sebastopol, CA: O'Reilly Media, 2010.

Levy, Steven. "The Final Days of Obama's Tech Surge." *Wired*, January 18, 2017. https://www.wired.com/2017/01/the-final-days-of-obamas-tech-surge/.

Libaw, Frieda B. "Information Handling in the Behavioral Sciences: Report of a First Convocation of a Conglomerate Clan." *American Behavioral Scientist* 10, no. 6 (February 1, 1967): 8–12. https://doi.org/10.1177/000276426701000603.

Libes, Don, and Sanford Ressler. *Life with UNIX: A Guide for Everyone*. Englewood Cliffs, NJ: Prentice Hall, 1989.

Lillington, Karlin. "PayPal Puts Dough in Your Palm." *Wired*, July 27, 1999. https://www.wired.com/1999/07/paypal-puts-dough-in-your-palm/.

Line, Maurice B. "Information Requirements in the Social Sciences: Some Preliminary Considerations." *Journal of Librarianship* 1, no. 1 (January 1, 1969): 1–19. https://doi.org/10.1177/096100066900100101.

Liu, Wendy. *Abolish Silicon Valley: How to Liberate Technology from Capitalism*. London: Repeater, 2020.

Long, Edward V. *The Intruders; the Invasion of Privacy by Government and Industry, by Edward V. Long. With a Foreword by Hubert H. Humphrey*. New York: Praeger, 1967.

Loughnane, Christopher, and William Aspray. "Rethinking the Call for a US National Data Center in the 1960s: Privacy, Social Science Research, and Data Fragmentation Viewed from the Perspective of Contemporary Archival Theory." *Information & Culture* 53, no. 2 (May 1, 2018): 203–242. https://doi.org/10.7560/IC53204.

Lubar, Steven. "'Do Not Fold, Spindle or Mutilate': A Cultural History of the Punch Card." *Journal of American Culture* 15, no. 4 (1992): 43–55. https://doi.org/10.1111/j.1542-734X.1992.1504_43.x.

Mackenzie, Adrian. *Wirelessness: Radical Empiricism in Network Cultures*. Cambridge, MA: MIT Press, 2010.

Mallinson, J. C. "Magnetic Tape Recording: Archival Considerations." In *[1990] Digest of Papers. Tenth IEEE Symposium on Mass Storage Systems@m_Crisis in Mass Storage*, Monterey, CA, 58–61. New York: IEEE, 1990. https://doi.org/10.1109/MASS.1990.113570.

Maron, Deborah, and Erin Carter. "'More Than What It Seems': How Critical Theory, Popular Engagement and Apps Like Tinder Can Help Us Reframe Metadata and Its Consequences." In *Proceedings of the International Conference on Dublin Core and Metadata Applications*, Washington, DC, October 26–29, 2017.

Marshall, Catherine C. "How People Manage Personal Information over a Lifetime." In *Personal Information Management*, edited by William P. Jones and Jaime Teevan, 57–75. Seattle: University of Washington Press, 2011.

Matienzo, María A. "To Hell with Good Intentions: Linked Data, Community and the Power to Name." *LITA Forum*, 2015. https://scholar.google.com/scholar?cluster=10216571721061765160&hl=en&oi=scholarr.

Mayernik, Matthew S. "Metadata." *Knowledge Organization* 47, no. 8 (2020): 696–713. https://doi.org/10.5771/0943-7444-2020-8-696.

Mayernik, Matthew S. "Metadata Accounts: Achieving Data and Evidence in Scientific Research." *Social Studies of Science* 49, no. 5 (October 1, 2019): 732–757. https://doi.org/10.1177/0306312719863494.

McCarthy, John. "Reminiscences on the History of Time-Sharing." *IEEE Annals of the History of Computing* 14, no. 1 (January 1, 1992): 19–24.

McKinney, Cait. *Information Activism: A Queer History of Lesbian Media Technologies*. Annotated ed. Durham, NC: Duke University Press Books, 2020.

Melo, Marijel (Maggie), Kimberly Hirsh, and Laura March. "Makerspaces in Libraries at U.S. Public Colleges and Universities: A Census." *Portal: Libraries and the Academy* 23, no. 1 (2023): 35–43. https://doi.org/10.1353/pla.2023.0007.

"Microtape: 555 Tape Transport/ 550 Tape Control/ Programs." Digital Equipment Corporation, 1964. Box 7, Folder 13. Digital Equipment Corporation records. http://www.computinghistory.org.uk/userdata/files/dec-microtape-555-tape-transport-550-tape-control-programs.pdf.

Miller, Arthur R. *The Assault on Privacy: Computers, Data Banks, and Dossiers*. Ann Arbor: University of Michigan Press, 1971.

Mirowski, Philip. *Science-Mart: Privatizing American Science*. Cambridge, MA: Harvard University Press, 2011.

Mirowski, Philip. "The Future(s) of Open Science." *Social Studies of Science* 48, no. 2 (April 1, 2018): 171–203. https://doi.org/10.1177/0306312718772086.

Mozilla. "Facebook and Google: This Is What an Effective Ad Archive API Looks Like." *The Mozilla Blog*, March 27, 2019. https://blog.mozilla.org/blog/2019/03/27/facebook-and-google-this-is-what-an-effective-ad-archive-api-looks-like.

Mozilla. "Open Letter: Facebook, Do Your Part Against Disinformation." *The Mozilla Blog*, February 11, 2019. https://blog.mozilla.org/blog/2019/02/11/open-letter-facebook-do-your-part-against-disinformation.

MuckRock. "Palantir—DCAS." April 14, 2016. https://www.muckrock.com/foi/new-york-city-17/palantir-dcas-25162/.

Munro, Iain. *Information Warfare in Business: Strategies of Control and Resistance in the Network Society*. London: Routledge, 2004. https://doi.org/10.4324/9780203449233.

Nakashima, Ellen. "NSA Has Halted a Counterterrorism Program Relying on Phone Records amid Doubts about Its Utility." *Washington Post*, March 6, 2019. https://www.washingtonpost.com/world/national-security/nsa-has-halted-a-counterterrorism-program-relying-on-phone-records-amid-doubts-about-its-utility/2019/03/05/f2d2793e-3f80-11e9-922c-64d6b7840b82_story.html.

Napoli, Philip M. "User Data as Public Resource: Implications for Social Media Regulation." *Policy & Internet* 11, no. 4 (December 2019): 439–459. https://doi.org/10.1002/poi3.216.

National Research Council. *Engaging Privacy and Information Technology in a Digital Age*. Edited by James Waldo, Herbert Lin, and Lynette I. Millett. Washington, DC: National Academies Press, 2007.

National Research Council, Division on Engineering and Physical Sciences, Computer Science and Telecommunications Board, and Committee on Responding to

Section 5(d) of Presidential Policy Directive 28: The Feasibility of Software to Provide Alternatives to Bulk Signals Intelligence Collection. *Bulk Collection of Signals Intelligence: Technical Options*. Washington, DC: National Academies Press, 2015.

National Security Agency/Central Security Service. "NSA Reports Data Deletion." June 28, 2018. https://www.nsa.gov/Press-Room/Press-Releases-Statements/Press-Release-View/Article/1618691/nsa-reports-data-deletion/http%3A%2F%2Fwww.nsa.gov%2FPress-Room%2FPress-Releases-Statements%2FPress-Release-View%2FArticle%2F1618691%2Fnsa-reports-data-deletion%2F.

Nielsen Research. "The State of Mobile Apps." *Nielsen*, June 2010. https://www.nielsen.com/insights/2010/the-state-of-mobile-apps-2/.

Nooney, Laine. *The Apple II Age: How the Computer Became Personal*. Chicago, IL: University of Chicago Press, 2023.

Nooney, Laine, Kevin Driscoll, and Kera Allen. "From Programming to Products: Softalk Magazine and the Rise of the Personal Computer User." *Information & Culture* 55, no. 2 (2020): 105–129.

NSA Civil Liberties and Privacy Office. "Transparency Report: THE USA FREEDOM Act Business Records FISA Implementation." January 15, 2016.

Office of Director of National Intelligence. "Statistical Transparency Report Regarding the Use of National Security Authorities: Calendar Year 2017." Office of Civil Liberties, Privacy, and Transparency, 2018.

Office of Director of National Intelligence. "Statistical Transparency Report Regarding the Use of National Security Authorities: Calendar Year 2018." Office of Civil Liberties, Privacy, and Transparency, 2019.

Office of the Press Secretary. "Fact Sheet: Data by the People, for the People—Eight Years of Progress Opening Government Data to Spur Innovation, Opportunity, & Economic Growth." Whitehouse.gov, September 28, 2016. https://obamawhitehouse.archives.gov/the-press-office/2016/09/28/fact-sheet-data-people-people-eight-years-progress-opening-government.

Ogul, Morris S. *Congress Oversees the Bureaucracy: Studies in Legislative Supervision / Morris S. Ogul*. Pittsburgh, PA: University of Pittsburgh Press, 1976.

Olson, Hope A. *The Power to Name: Locating the Limits of Subject Representation in Libraries*. Dordrecht: Springer, 2011.

Onions, C. T., G. W. S. Friedrichsen, and R. W. Burchfield. *The Oxford Dictionary of English Etymology*. Oxford: Clarendon Press, 1979.

Ornstein, Severo. *Computing in the Middle Ages: A View from the Trenches 1955–1983*. Bloomington, IN: First Books, 2002.

Osgood, Kenneth, Andrew K. Frank, Andrew L. Johns, and Kenneth Osgood, eds. "'We Need to Get a Better Story to the American People': LBJ, the Progress

Campaign, and the Vietnam War on Television." In *Selling War in a Media Age: The Presidency and Public Opinion in the American Century*. Gainesville: University Press of Florida, 2010. https://doi.org/10.5744/florida/9780813034669.003.0008.

Packingham v. North Carolina, No. 15–1194. Supreme Court of the United States, June 19, 2017.

"Palm Position Paper." Palm Computing Inc., 1993. Box 1, Folder 1. Computer History Museum, Jeff Hawkins papers.

PalmSource. *Zen of Palm: Designing Products for Palm OS*. Sunnyvale, CA: PalmSource, 2003. http://archive.org/details/zen-of-palm.

Pasquale, Frank. "From Territorial to Functional Sovereignty: The Case of Amazon," *Law and Political Economy* 6 (2017), https://lpeproject.org/blog/from-territorial-to-functional-sovereignty-the-case-of-amazon/.

Pasquale, Frank. *The Black Box Society: The Secret Algorithms That Control Money and Information*. Reprint ed. Cambridge, MA: Harvard University Press, 2016.

PDA News. "Developer Forum: Interview with Jeff Hawkins." March 1994. Box 1, Folder 1. Computer History Museum, Jeff Hawkins papers.

PDA News. "Simon-Ized: Part Cellular Phone, Part Touch Sensitive Key Pad, BellSouth's Simon May Revolutionize the Industry." March 1994, Vol. 1, Number 3 edition.

PDA News. "Zoomer Package Both Functional and Promising." March 1994. Box 1, Folder 1. Computer History Museum, Jeff Hawkins papers.

"PDP-6 Time Sharing Software." Digital Equipment Corporation, 1965. Box 60, Folder 5. Digital Equipment Corporation records. http://bitsavers.org/pdf/dec/pdp6/PDP-6_TimsharingBroch.pdf.

"PDP-42-2: Preliminary Microtape File System." PDP-1 Computer. Cambridge, MA: Electrical Engineering Department, Massachusetts Institute of Technology, February 6, 1970. Digital Equipment Corporation records.

Pemberton, David M. "'Blisters on My Heels, Corns on My Toes': Taking the 1930 Census of Population." *National Archives* 34, no. 4 (2002). https://www.archives.gov/publications/prologue/2002/winter/1930-census.html.

Plavchan, Ronald. "The International Scene: News and Abstracts." *The American Archivist* 42, no. 2 (April 1, 1979): 230. https://doi.org/10.17723/aarc.42.2.r367725p283xt569.

Poirier, Lindsay. "Classification as Catachresis: Double Binds of Representing Difference with Semiotic Infrastructure." *Canadian Journal of Communication* 44, no. 3 (2019): 361–371.

Poirier, Lindsay. *Knowledge Representation in Scruffy Worlds: An Ethnography of Semiotic Infrastructure Design Work*. Troy, NY: Rensselaer Polytechnic Institute, 2018.

Porter, Jon. "Twitter Announces New API Pricing, Posing a Challenge for Small Developers." *The Verge*, March 30, 2023. https://www.theverge.com/2023/3/30/23662832/twitter-api-tiers-free-bot-novelty-accounts-basic-enterprice-monthly-price.

Porter, Theodore M. *The Rise of Statistical Thinking, 1820–1900*. New ed. Princeton, NJ: Princeton University Press, 2020.

"Privacy and Efficient Government: Proposals for a National Data Center." *Harvard Law Review* 82, no. 2 (December 1968): 400. https://doi.org/10.2307/1339231.

"Programmed Data Processor-1 Manual." Digital Equipment Corporation, 1961. 102664957. Computer History Museum. https://www.computerhistory.org/pdp-1/50fc0d6d57d4a4d67885984808c8d847/.

"Public Statement from the Co-Chairs and European Advisory Committee of Social Science One." December 11, 2019. https://socialscience.one/blog/public-statement-european-advisory-committee-social-science-one.

Purcell, Kristen, Roger Entner, and Nichole Henderson. "The Rise of Apps Culture." *Pew Research Center: Internet, Science & Tech* (blog), September 14, 2010. https://www.pewresearch.org/internet/2010/09/14/the-rise-of-apps-culture/.

Pyke, Thomas N. "Time-Shared Computer Systems." *Advances in Computers* 8 (1967): 1–45. https://doi.org/10.1016/S0065-2458(08)60693-2.

Raley, Rita. "Dataveillance and Countervailance." In *"Raw Data" Is an Oxymoron*, edited by Lisa Gitelman, 121–146. Cambridge, MA: MIT Press, 2013.

Ramey, James W. "Urban Data Banks and the Rights of the Individual." *Socio-Economic Planning Sciences* 1, no. 3 (July 1, 1968): 327–333. https://doi.org/10.1016/0038-0121(68)90019-0.

Rankin, Joy Lisi. *A People's History of Computing in the United States*, illustrated ed. Cambridge, MA: Harvard University Press, 2018.

Regan, Priscilla M. *Legislating Privacy: Technology, Social Values, and Public Policy*. Chapel Hill: University of North Carolina Press, 1995.

Ribak, Rivka. "Translating Privacy: Developer Cultures in the Global World of Practice." *Information, Communication & Society* 22, no. 6 (May 12, 2019): 838–853. https://doi.org/10.1080/1369118X.2019.1577475.

Ridener, John. *From Polders to Postmodernism: A Concise History of Archival Theory*. Duluth, MN: Litwin Books, 2009.

Rieke, A., and M. Bogen. "Leveling the Platform: Real Transparency for Paid Messages on Facebook." *Upturn*, May 9, 2018. https://www.upturn.org/work/leveling-the-platform.

Ritchie, D. M. "The UNIX System: The Evolution of the UNIX Time-Sharing System." *AT&T Bell Laboratories Technical Journal* 63, no. 8 (1984): 1577–1593. https://doi.org/10.1002/j.1538-7305.1984.tb00054.x.

Robertson, Craig. "The Filing Cabinet." *Places Journal*, May 11, 2021. https://doi.org/10.22269/210511.

Robertson, Craig. *The Filing Cabinet: A Vertical History of Information*. Minneapolis: University of Minnesota Press, 2021.

Robertson, Nan. "Data-Center Aims Scored in Inquiry: Advocates Hear Legislators Challenge Their System." *New York Times*, 1966.

Ruggles, Richard, Richard Miller, Edwin Kuh, Stanley Lebergott, Guy Orcutt, and Joseph Penchman. "Report of the Committee on the Preservation and Use of Economic Data (1965)." Accessed November 16, 2021. http://archive.org/details/ReportOfTheCommitteeOnThePreservationAndUseOfEconomicData1965.

Rule, James B., *The Politics of Privacy: Planning for Personal Data Systems as Powerful Technologies*. New York: Elsevier, 1980.

Rusbridger, Alan. "The Snowden Leaks and the Public." *New York Review of Books*, November 21, 2013.

Russell, Andrew L. *Open Standards and the Digital Age: History, Ideology, and Networks*. New York: Cambridge University Press, 2014.

Ryu, Min Ho, Junghwan Kim, and Seongcheol Kim. "Factors Affecting Application Developers' Loyalty to Mobile Platforms." *Computers in Human Behavior* 40 (November 1, 2014): 78–85. https://doi.org/10.1016/j.chb.2014.08.001.

Sadowski, Jathan. *Too Smart: How Digital Capitalism Is Extracting Data, Controlling Our Lives, and Taking Over the World*. Cambridge, MA: MIT Press, 2020.

Salus, Peter. *A Quarter Century of UNIX*. Reading, MA: Addison-Wesley, 1994. https://dl.acm.org/doi/abs/10.5555/191771.

Sawyer, Jack, and Howard Schechter. "Computers, Privacy, and the National Data Center: The Responsibility of Social Scientists." *American Psychologist* 23, no. 11 (November 1968): 810–818. https://doi.org/10.1037/h0026719.

Schmandt-Besserat, Denise. *How Writing Came About*. Abridged ed. Austin: University of Texas Press, 1997.

Schudson, Michael. *The Rise of the Right to Know: Politics and the Culture of Transparency, 1945–1975*. Cambridge, MA: Belknap Press, 2018.

Sellen, Abigail J., and Richard Harper. *The Myth of the Paperless Office*. Cambridge, MA: MIT Press, 2002.

Senate Committee on the Judiciary, Subcommittee on Constitutional Rights. "Federal Data Banks and Constitutional Rights." 1971.

Senior Advisory Group Panel. "Declassified Report on Commercially Available Information." Office of the Director of National Intelligence. Washington, DC: United States Office of the Director of National Intelligence, January 27,

2022. https://www.dni.gov/index.php/newsroom/reports-publications/reports-publications-2023/3700-odni-senior-advisory-group-panel-declassified-report-on-commercially-available-information.

Shilton, Katie. "Values Levers: Building Ethics into Design." *Science, Technology, & Human Values* 38, no. 3 (April 23, 2012): 374–397. https://doi.org/10.1177/0162243912436985.

Shilton, Katie, and Daniel Greene. "Linking Platforms, Practices, and Developer Ethics: Levers for Privacy Discourse in Mobile Application Development." *Journal of Business Ethics* 155, no. 1 (March 1, 2019): 131–146. https://doi.org/10.1007/s10551-017-3504-8.

Siewiorek, Daniel, Gordon Bell, and Allen Newell. *Computer Structures: Principles and Examples*. New York: McGraw Hill, 1982. https://dl.acm.org/doi/abs/10.5555/578056.

Silverman, Craig. "Facebook Said It Would Give Detailed Data to Academics. They're Still Waiting." *BuzzFeed News*, August 23, 2019. https://www.buzzfeednews.com/article/craigsilverman/slow-facebook.

Skeat, Walter W. *An Etymological Dictionary of the English Language*. New ed., rev. enl. Oxford: Clarendon Press, 1924.

Snap Inc. "2019 Snap Research Fellowship & Scholarship." Snap Inc., 2019. https://snapresearchfs.splashthat.com.

Snap Inc. "Snap Political and Advocacy Ads Library—Business Center, Frequently Asked Questions." Snap Chat Business Center, 2020. https://businesshelp.snapchat.com/en-US/article/political-ads-library.

Social Science One. "Overview." *Social Science One*, 2018. https://socialscience.one/overview.

Sofia, Zoë. "Container Technologies." *Hypatia* 15, no. 2 (April 2000): 181–201. https://doi.org/10.1111/j.1527-2001.2000.tb00322.x.

Srnicek, Nick. *Platform Capitalism*. 1st ed. Cambridge: Polity, 2016.

Steinkeller, Piotr. "Archival Practices at Babylonia in the Third Millennium." In *Ancient Archives and Archival Traditions: Concepts of Record-Keeping*, edited by Maria Brosius, 1–16. Oxford Studies in Ancient Documents. Oxford: Oxford University Press, 2003.

Sterne, Jonathan. *MP3: The Meaning of a Format*. Durham, NC: Duke University Press Books, 2012.

Sutherland, Tonia. "Making a Killing: On Race, Ritual, and (Re)Membering in Digital Culture." *Preservation, Digital Technology & Culture* 46, no. 1 (2017): 32–40. https://doi.org/10.1515/pdtc-2017-0025.

Tang, John, Jed R. Brubaker, Catherine C. Marshall, and Cathy Marshall. "What Do You See in the Cloud? Understanding the Cloud-Based User Experience through Practices." Microsoft, September 2, 2013. https://www.microsoft.com/en-us/research/publication/what-do-you-see-in-the-cloud-understanding-the-cloud-based-user-experience-through-practices/.

Taylor, Emily, and Stacie Hoffman. "Industry Responses to Computational Propaganda and Social Media Manipulation." Working Paper. Oxford: Project on Computational Propaganda, 2019. https://comprop.oii.ox.ac.uk/research/industry-responses/.

"Text—H.R.2048—114th Congress (2015–2016): USA FREEDOM Act of 2015." June 2, 2015. http://www.congress.gov/.

"The Palm Personal Information Manager." Palm, 1993. Box 1, Folder 1. Computer History Museum, Jeff Hawkins papers.

The White House. "FACT SHEET: United States and European Commission Announce Trans-Atlantic Data Privacy Framework." The White House, March 25, 2022. https://www.whitehouse.gov/briefing-room/statements-releases/2022/03/25/fact-sheet-united-states-and-european-commission-announce-trans-atlantic-data-privacy-framework/.

The White House. "Statement by the President." The White House, June 7, 2013. https://obamawhitehouse.archives.gov/the-press-office/2013/06/07/statement-president.

Thibodeau, Kenneth. "Overview of Technological Approaches to Digital Preservation and Challenges in Coming Years." Washington, DC: Council on Library and Information Resources, 1997.

Thylstrup, Nanna Bonde. *The Politics of Mass Digitization*. Cambridge, MA: MIT Press, 2019.

Tkacz, Nathaniel. *Being with Data: The Dashboarding of Everyday Life*. 1st ed. Cambridge: Polity, 2022.

Tobin, Ariana, and Jeremy B. Merrill,. "Facebook Moves to Block Ad Transparency Tools—Including Ours." *ProPublica*, January 28, 2019. https://www.propublica.org/article/facebook-blocks-ad-transparency-tools.

Toomey, Warren. "The Strange Birth and Long Life of Unix." *IEEE Spectrum*, November 28, 2011. https://spectrum.ieee.org/the-strange-birth-and-long-life-of-unix.

Trace, Ciaran B. "On or Off the Record? Notions of Value in the Archive." In *Currents of Archival Thinking*, edited by Terry Eastwood and Heather MacNeil, 47–68. Santa Barbara, CA: Libraries Unlimited, 2010. http://archive.org/details/currentsofarchiv0000unse_d0m0.

Trace, Ciaran B. "What Is Recorded Is Never Simply 'What Happened': Record Keeping in Modern Organizational Culture." *Archival Science* 2, no. 1 (March 1, 2002): 137–159. https://doi.org/10.1007/BF02435634.

Tromble, Rebekah. "Where Have All the Data Gone? A Critical Reflection on Academic Digital Research in the Post-API Age." *Social Media + Society* 7, no. 1 (January 1, 2021): 2056305121988929. https://doi.org/10.1177/2056305121988929.

Turner, Fred. *From Counterculture to Cyberculture: Stewart Brand, the Whole Earth Network, and the Rise of Digital Utopianism*. Chicago, IL: University of Chicago Press, 2006.

Twitter. "Disclosing State-Linked Information Operations We've Removed." Twitter (blog), December 2, 2021. https://blog.twitter.com/en_us/topics/company/2021/disclosing-state-linked-information-operations-we-ve-removed.

Twitter Inc. "Updating Our Advertising Policies on State Media." Twitter (blog), August 19, 2019. https://blog.twitter.com/en_us/topics/company/2019/advertising_policies_on_state_media.html.

Twitter Safety. "Disclosing New Data to Our Archive of Information Operations." Twitter (blog), September 20, 2019. https://blog.twitter.com/en_us/topics/company/2019/info-ops-disclosure-data-september-2019.html.

Twitter Safety. "New Disclosures to Our Archive of State-Backed Information Operations." Twitter (blog), December 20, 2019. https://blog.twitter.com/en_us/topics/company/2019/new-disclosures-to-our-archive-of-state-backed-information-operations.html.

US Department of Health, Education and Welfare, Secretary's Advisory Committee on Automated Personal Data Systems. "Records, Computers, and the Rights of Citizens." Washington, DC: Government Printing Office, 1973.

US House Committee on Government Operations, Special Subcommittee on Invasion of Privacy. "The Computer and Invasion of Privacy: Hearings." 89th Cong., July 26–28, 1966. Washington, DC, 1966.

US Senate Committee on the Judiciary, Subcommittee on Administrative Practice and Procedure. "Invasions of Privacy (Government Agencies): Hearings." 89th Cong., 2nd sess., pt. 5, March 23–30, June 7–9, 14, 16, 1966. Washington, DC, 1966.

US Senate Committee on the Judiciary, Subcommittee on Administrative Practice and Procedure. "Government Dossier (Survey of Information Contained in Government Files)." 90th Cong., 1st sess., 1967.

Vaidhyanathan, Siva. *Antisocial Media: How Facebook Disconnects Us and Undermines Democracy*. New York: Oxford University Press, 2018.

Vander Noot, T. J. "Computers, Social Science Statisticians, and the A.S.A." *The American Statistician* 22, no. 3 (1968): 20–22. https://doi.org/10.2307/2681798.

van Dijck, José. "Datafication, Dataism and Dataveillance: Big Data between Scientific Paradigm and Ideology." *Surveillance & Society* 12, no. 2 (May 9, 2014): 197–208. https://doi.org/10.24908/ss.v12i2.4776.

van Dijck, José. *The Culture of Connectivity: A Critical History of Social Media*. 1st ed. Oxford: Oxford University Press, 2013.

van Dijck, José, Thomas Poell, and Martijn de Waal. *The Platform Society: Public Values in a Connective World*. New York, NY: Oxford University Press, 2018.

Viljoen, Salomé. "A Relational Theory of Data Governance." *The Yale Law Journal* 131, no. 2 (November 2021): 573–654.

Villa-Nicholas, Melissa. *Data Borders: How Silicon Valley is Building an Industry Around Immigrants*. Oakland, CA: University of California Press, 2023.

Vismann, Cornelia. *Files: Law and Media Technology*. Translated by Geoffrey Winthrop-Young. Stanford, CA: Stanford University Press, 2008.

Volz, Dustin, and Warren P. Strobel. "NSA Recommends Dropping Phone-Surveillance Program." *Wall Street Journal*. Accessed December 11, 2022. https://www.wsj.com/articles/nsa-recommends-dropping-phone-surveillance-program-11556138247.

Walker, Shawn. "The Complexity of Collecting Digital and Social Media Data in Ephemeral Contexts." PhD thesis, University of Washington, 2017. https://digital.lib.washington.edu:443/researchworks/handle/1773/40612.

Warner, Malcolm, and Michael Graham Stone. *The Data Bank Society: Organizations, Computers and Social Freedom*. London: Allen & Unwin, 1970.

Wesolowski, Amy, Caroline O. Buckee, Linus Bengtsson, Erik Wetter, Xin Lu, and Andrew J. Tatem. "Commentary: Containing the Ebola Outbreak—The Potential and Challenge of Mobile Network Data." *PLoS Currents* 6 (September 29, 2014): ecurrents.outbreaks.0177e7fcf52217b8b634376e2f3efc5e. https://doi.org/10.1371/currents.outbreaks.0177e7fcf52217b8b634376e2f3efc5e.

Westin, Alan F. *Privacy and Freedom*. 1st ed. New York: Atheneum, 1967.

Westin, Alan F., and Michael A. Baker. "Databanks in a Free Society: Computers, Record-Keeping and Privacy." New York, NY: Quadrangle Books, 1972.

WhatsApp. "WhatsApp Research Awards for Social Science and Misinformation." WhatsApp.com, 2018. https://www.whatsapp.com/research/awards.

Wiebe, Robert H. *The Search for Order, 1877–1920*. 1st ed. New York: Hill and Wang, 1966.

Wight, A. S. "The EMAS Archiving Program." *The Computer Journal* 18, no. 2 (January 1, 1975): 131–134. https://doi.org/10.1093/comjnl/18.2.131.

Wilkes, Mary Allen, and W. A. Clark. *Programming the LINC*. St. Louis, MO: Computer Research Laboratory, Washington University, 1965.

Woolsey, T. D. "Data Banks Are Not the Answer: A Statistician's Viewpoint." *American Journal of Public Health and the Nation's Health* 60, no. 10 (October 1, 1970): 1991–1995. https://doi.org/10.2105/AJPH.60.10.1991.

Yates, JoAnne. *Control Through Communication: The Rise of System in American Management*. Baltimore, MD: Johns Hopkins University Press, 1993.

Yates, John E. "A Time Sharing System for the PDP-1 Computer." Scientific Report No. 5. Electronic Systems Laboratory. Cambridge, MA: Massachusetts Institute of Technology, June 1962. Box 4, Folder 1. Computer History Museum, PDP-1 records.

Yelp. "Yelp Open Dataset: An All-Purpose Dataset for Learning." *Yelp*, 2020. https://www.yelp.com/dataset.

Zimmer, Michael. "The Twitter Archive at the Library of Congress: Challenges for Information Practice and Information Policy." *First Monday* 20, no. 7 (June 21, 2015). http://firstmonday.org/ojs/index.php/fm/article/view/5619.

Zuboff, Shoshana. *The Age of Surveillance Capitalism: The Fight for a Human Future at the New Frontier of Power*. New York: PublicAffairs, 2019.

INDEX

Page numbers followed by an *f* denote a figure.

Publisher contact:
The MIT Press
Massachusetts Institute of Technology
77 Massachusetts Avenue, Cambridge, MA 02139
mitpress.mit.edu

EU Authorised Representative:
Easy Access System Europe, Mustamäe tee 50,
10621 Tallinn, Estonia
gpsr.requests@easproject.com

Printed by Integrated Books International,
United States of America